MEMOIRS
of the
American Mathematical Society

Number 452

Derivates of
Interval Functions

Brian S. Thomson

Forsyth Library
Fort Hays State University

September 1991 • Volume 93 • Number 452 (first of 3 numbers) • ISSN 0065-9266

American Mathematical Society
Providence, Rhode Island

1980 *Mathematics Subject Classification* (1985 *Revision*).
Primary 26A21, 26A24.

Library of Congress Cataloging-in-Publication Data

Thomson, Brian S., 1941–
   Derivates of interval functions/Brian S. Thomson.
   p. cm. – (Memoirs of the American Mathematical Society, ISSN 0065-9266; no. 452)
   Includes bibliographical references and index.
   ISBN 0-8218-2503-8
   1. Interval functions.   2. Differentiable functions.   I. American Mathematical Society.   II. Title.   III. Series.
QA3.A57   no. 452
[QA331.5]
510 s–dc20   91-22745
[515′.8]   CIP

Subscriptions and orders for publications of the American Mathematical Society should be addressed to American Mathematical Society, Box 1571, Annex Station, Providence, RI 02901-1571. *All orders must be accompanied by payment.* Other correspondence should be addressed to Box 6248, Providence, RI 02940-6248.

SUBSCRIPTION INFORMATION. The 1991 subscription begins with Number 438 and consists of six mailings, each containing one or more numbers. Subscription prices for 1991 are $270 list, $216 institutional member. A late charge of 10% of the subscription price will be imposed on orders received from nonmembers after January 1 of the subscription year. Subscribers outside the United States and India must pay a postage surcharge of $25; subscribers in India must pay a postage surcharge of $43. Expedited delivery to destinations in North America $30; elsewhere $82. Each number may be ordered separately; *please specify number* when ordering an individual number. For prices and titles of recently released numbers, see the New Publications sections of the NOTICES of the American Mathematical Society.

BACK NUMBER INFORMATION. For back issues see the AMS Catalogue of Publications.

MEMOIRS of the American Mathematical Society (ISSN 0065-9266) is published bimonthly (each volume consisting usually of more than one number) by the American Mathematical Society at 201 Charles Street, Providence, Rhode Island 02904-2213. Second Class postage paid at Providence, Rhode Island 02940-6248. Postmaster: Send address changes to Memoirs of the American Mathematical Society, American Mathematical Society, Box 6248, Providence, RI 02940-6248.

   COPYING AND REPRINTING. Individual readers of this publication, and nonprofit libraries acting for them, are permitted to make fair use of the material, such as to copy an article for use in teaching or research. Permission is granted to quote brief passages from this publication in reviews, provided the customary acknowledgment of the source is given.
   Republication, systematic copying, or multiple reproduction of any material in this publication (including abstracts) is permitted only under license from the American Mathematical Society. Requests for such permission should be addressed to the Manager of Editorial Services, American Mathematical Society, P.O. Box 6248, Providence, Rhode Island 02940-6248.
   The owner consents to copying beyond that permitted by Sections 107 or 108 of the U.S. Copyright Law, provided that a fee of $1.00 plus $.25 per page for each copy be paid directly to the Copyright Clearance Center, Inc., 27 Congress Street, Salem, Massachusetts 01970. When paying this fee please use the code 0065-9266/91 to refer to this publication. This consent does not extend to other kinds of copying, such as copying for general distribution, for advertising or promotional purposes, for creating new collective works, or for resale.

Copyright © 1991, American Mathematical Society. All rights reserved.
Printed in the United States of America.
The paper used in this book is acid-free and falls within the guidelines
established to ensure permanence and durability. ∞

10 9 8 7 6 5 4 3 2 1   95 94 93 92 91

Contents

CONTENTS

ABSTRACT [1]

This report is motivated by a study of Rogers and Taylor characterizing those interval functions which are absolutely continuous with respect to the s–dimensional Hausdorff measure. This problem leads naturally to an investigation of Lipschitz numbers

$$D^s(f, x) = \limsup_{y,z \to x,\, y < x < z} \frac{f(z) - f(y)}{(z - y)^s}$$

and to s–dimensional integrals roughly of the form

$$\int_a^b f(x)\,dx^s = \lim \sum_{i=1}^{n} f(\xi_i)(x_i - x_{i-1})^s.$$

The exposition is presented in the setting of interval functions on the real line and the differentiation, measure-theoretic and variational properties are developed. Applications are given to the Hausdorff and packing measures as well as to the classical differentiation theory of real functions.

Received by the editor October 5, 1988. Received in revised form November 5, 1990.

[1] 1980 *Mathematics Subject Classification* (1985 *Revision*). Primary 26A21, 26A24.

1   Introduction

In the study of the derivation properties of interval functions there are certain arguments that reappear in many settings. It is our pupose here to present a unified approach to some of these techniques. In order to focus our attention we will present this material with a specific view to examining an interesting and important study of Rogers and Taylor [16] characterizing those interval functions which are in a sense absolutely continuous with respect to the s–dimensional Hausdorff measure. We follow Rogers [15] in confining our attention to the one-dimensional case although many of the arguments can be used in higher dimensions where richer geometry and larger choice of differentiation bases might obscure the natural simplicity of the ideas .

The problem is to determine the nature of continuous, nondecreasing functions f on the interval $[0, 1]$ whose Lebesgue-Stieltjes measures λ_f are absolutely continuous with respect to the s–dimensional Hausdorff measure on $[0, 1]$. This problem leads naturally to an investigation of derivates of the form

$$D^s(f, x) = \limsup_{y, z \to x, \, y < x < z} \frac{f(z) - f(y)}{(z - y)^s} \qquad (1)$$

which Besicovich [1] has called "Lipschitz numbers".

The discussion will be intimately related to the notion of a generalized Riemann integral as defined by Henstock [7] and Kurzweil [11]. Recall that the classical Riemann integral defined as a limit of Riemann sums

$$\int_a^b f(x)dx = \lim \sum_{i=1}^n f(\xi_i)(x_i - x_{i-1}) \qquad (2)$$

has been shown to allow greater flexibility than appears at a first study. By altering the definition of the limit in equation (2) one can obtain integrals that express exactly the improper Riemann integral, the Lebesgue integral, the Denjoy-Perron integral, the Denjoy-Khintchine integral, the approximate Perron and the approximate symmetric Perron integrals. The derivation theory for these integrals is particularly immediate for the same concept (what we call a "full covering relation" in Section 2) expresses both the integral and derivative, revealing a single unifying connection between these two classical ideas.

In connection with the problem of Rogers and Taylor one might expect a similar approach to relate Hausdorff measures, Lipschitz numbers and absolute continuity (the ingredients of their problem). Thus one might try for an integral of the form

$$\int_a^b f(x)dx^s = \lim \sum_{i=1}^n f(\xi_i)(x_i - x_{i-1})^s \qquad (3)$$

for $0 < s < 1$ under an appropriate interpretation of the limit process.

The integral defined in equation (3) cannot be defined in exactly the same manner as that in (2); thus the integration theory that is available for such integrals does not immediately apply. Nonetheless there are a number of appropriate interpretations of (3) that appear to be useful and which have connections with the theory of Hausdorff measures, illuminating some of the notions that arise there. In particular the integral of Besicovitch in [1] permits such a realization.

We begin with a general theory for the derivation and integration of interval functions, centered on the study of limits of the form

$$\limsup_{|I| \to 0, x \in I^0} h(I)/k(I),$$

where the limit involves approach to x by intervals always containing x as an interior point. Essentially then we have a study of this particular differentiation basis. Other differentiation bases could certainly have been chosen. This one has particularly smooth properties in many regards and is directed at the study of Lipschitz numbers. General notions of limit, derivative, continuity, variation, integral, absolute continuity and singularity are studied in this setting. The main theme throughout could be said to be a systematization of Vitali arguments.

We confine our applications to the study of functions of a real variable. In Section 6 we discuss some classical measures on the real line. In particular we show how the Lebesgue measure, the Lebesgue-Stieltjes measures and the Hausdorff measures on the line can be studied in this setting. We then turn to some classical properties of real functions, especially derivation and variation. The familiar properties of monotonic functions in Section 7.1 come as applications of the general differentiation results for interval functions. The material of Denjoy, Lusin and Saks on the structure of VBG_* functions is developed in Section 7.2; a number of the characterizations given appear to be new.

Sections 7.4 and 7.5 return to the problem of Rogers and Taylor – determining the nature of functions that are absolutely continuous with repect to the Hausdorff measures. Note that their study focuses on functions of bounded variation (or more properly their Lebesgue-Stieltjes measures) whereas here the methods apply directly to functions that are VBG$_*$ on a set. Thus we are able to state their results in a bit more generality, but more interestingly as an application of techniques that apply to general interval functions. At the same time we are able to prove a parallel characterization of the functions that are absolutely continuous with respect to the s–dimensional packing measure of Tricot.

The only background required of the reader is a familiarity with such standard parts of analysis as may be found in the well-known treatise of Saks [17]. In the discussion of Hausdorff measures in Section 6.3 it is assumed that the reader is familiar with the rudiments of that subject and in particular the density theorems for those measures. The intention of this memoir is to introduce a new perspective on certain applications of the Vitali covering theorem. While specialists will see how this could be applied in a number of different settings even the general reader may find the presentation of the classical differentiation theory of real functions unusual and, perhaps, interesting.

2   Covering relations

By a *covering relation* is meant a collection of pairs (I, x) where I is a closed interval and $x \in I$. The language is taken from Federer [5, p. 151]. A number of the constructions in classical analysis may be expressed using the notion of a covering relation and many arguments can be simplified by an appeal to this language.

We illustrate with an example. If $F' = f$ everywhere how might properties of F be recovered from a knowledge only of the function f? Of course if f is integrable then the solution is transparent: the integral $\int_a^x f(t)\, dt$ recovers F up to an additive constant and $\int_E |f(t)|\, dt$ recovers the variation of F on a measurable set E. If f is not integrable in either the Lebesgue or improper Riemann senses it is less clear how to proceed. Such a function f would have a dense set of intervals on which it is Lebesgue integrable; by using this fact together with deeper properties of derivatives Denjoy in 1912 was able to develop a countable, but transfinite, sequence of extensions of the Lebesgue integral which solve this problem.

A simpler analysis is available directly from the meaning of the assertion $F'(x) = f(x)$. For any positive number ϵ the collection

$$\beta = \left\{ ([y,z],x) : \left| \frac{F(z) - F(y)}{z - y} - f(x) \right| < \epsilon \right\} \qquad (4)$$

has an obvious local property at each point. Specifically at each point x and for every sufficiently small interval I with $x \in I$ the pair (I,x) belongs to β. One can view this as a relation connecting each point with many intervals.

For the problem at hand one can evidently recover the increment $F(b) - F(a)$ within an error of $\epsilon(b-a)$ by forming the sum

$$\sum_{i=1}^{n} f(\xi_i)(x_i - x_{i-1})$$

taken over a partition $a = x_0 < x_1 < \ldots < x_n = b$ for which each $([x_{i-1}, x_i], \xi_i) \in \beta$. This is the basis for the theory of generalized Riemann integrals developed by Henstock [7] and Kurzweil [11], and the argument of the previous sentence is almost completely a proof of the fundamental theorem of the calculus in this setting.

Subtle properties of covering relations may be explored geometrically and applied in a unified manner to give insights into the structure of real functions. The idea in essence is due to Vitali although the language is developed here somewhat differently. The original application of Vitali arises from the observation that from certain covering relations β that might hold for a set E of real numbers one may extract a sequence

$$\{(I_1, x_1), (I_2, x_2), \ldots, (I_n, x_n)\}$$

where the intervals $\{I_i\}$ are nonoverlapping and provide an approximation to the outer Lebesgue measure $\lambda(E)$ of the set E:

$$\lambda(E) \approx \sum_{i=1}^{\infty} |I_i|.$$

Evidently this approximation can be used to connect measure properties of certain sets with differentiation properties of functions. In particular such a technique can be used to give a transparent proof of the fact, just mentioned, that $\int_E |f(t)|\, dt$ recovers the variation of F on a measurable set E. More importantly it permits generalizations to situations where the Lebesgue integral is not directly applicable.

Thus the separate notions of differentiation, measure and integration are all linked by the notion of a covering relation. Here we continue this theme and organize it in a language that allows the generalizations to develop.

2.1 Basic language of covering relations

We introduce the general notion of a covering relation and develop the language needed for our discussion of these relations.

2.1 DEFINITION. A *covering relation* on a set of real numbers E is a collection of pairs (I, x) where I is a closed interval and $x \in I \cap E$.

If the collection of all closed intervals is called $\mathcal{I}$ then a covering relation is merely a subset β of the product $\mathcal{I} \times \mathbb{R}$. We prefer lower case greek letters for covering relations, usually employing α, β, γ or π.

2.2 DEFINITION. If β is a covering relation and E is a set of real numbers then $\beta(E)$, $\beta[E]$, and $\sigma(\beta)$ denote the following sets:

1. $\beta(E) = \{(I, x) \in \beta : I \subset E\}$,

2. $\beta[E] = \{(I, x) \in \beta : x \in E\}$.

3. $\sigma(\beta) = \bigcup_{(I,x)\in\beta} I$.

The expressions $\beta(E)$ and $\beta[E]$ are also covering relations and are subsets of β. The passage to $\beta(E)$ and $\beta[E]$ from β is a common device in derivation theory. In some settings (eg. in [6, p.12]) $\beta(E)$ is called a "pruning" of β; the language is meant to indicate that some inessential members of β have been removed. The most common pruning of a relation β on a set E will be to form $\beta(G)$ for an open set G that contains E.

2.3 DEFINITION. A *packing* is a covering relation π with the property that for distinct pairs (I_1, x_1) and (I_2, x_2) belonging to π the intervals I_1 and I_2 do not overlap.

Evidently a packing is either finite or countably infinite. Using the language of Henstock [7], we call a finite packing π a *division* of an interval $[a, b]$ if $\sigma(\pi) = [a, b]$. Some authors call this a partition; throughout we shall employ this term in its usual set-theoretic sense.

2.2 Full and fine covering relations

Most of the covering relations that one encounters in analysis have certain local properties. The covering relations and covering lemmas that are our central concern are those that arise naturally in discussions of limits or derivatives. We now introduce the notion of a full covering relation and its dual notion, that of a fine covering relation. They will play a central role in all of our studies. Since our focus is on limits of the form

$$\lim_{y,z \to x, \ y < x < z} \frac{|f(y) - f(z)|}{(y - z)^s},$$

we require the following special form of the covering relations. Note in particular that the pairs (I, x) in the relations always have x an interior point of the closed interval I.

2.4 DEFINITION. A covering relation β is said to be *full* at a point x provided that there exists a $\delta > 0$ so that $([y, z], x) \in \beta$ for every $y < x < z$ with $0 < z - y < \delta$. Such a relation is said to be a *full covering relation* on a set E if it is full at each point of E.

We give also a version parallel to this; this is essentially the notion of a Vitali covering.

2.5 DEFINITION. A covering relation β is said to be *fine* at a point x provided that for every $\epsilon > 0$ there is a pair $([y, z], x) \in \beta$, with $y < x < z$ and $0 < z - y < \epsilon$. Again β is a *fine covering relation* on a set E if it is fine at each point of E.

The relation between full and fine covering relations can be expressed by the following simple observation. Let β be a covering relation and let

$$\gamma = (\mathcal{I} \times \mathbb{R}) \setminus \beta.$$

Then γ is full at a point x if and only if β is not fine at x. This may be thought of as a duality (or more properly a negative duality) and can be made the basis for a genuine dual relationship between pairs of differentiation bases (as in [21] and [22]); we shall not need any such formal apparatus. Nonetheless most statements for upper and lower derivates will appear in dual pairs that correspond to the notions of full and fine covering relations and measures will appear in dual pairs that are similarly related.

By a *covering lemma* we mean an assertion that from one or several covering relations some subcovering may be constructed. The model is the classical Vitali covering theorem. Expressed in our language this theorem asserts that if β is a fine covering relation on the set E then there is a collection

$$\gamma = \{(I_i, x_i) : i = 1, 2, \ldots n\}$$

contained in β for which the intervals I_i are pairwise disjoint and the sum $\sum_{(I,x) \in \gamma} |I|$ approximates the measure of E arbitrarily closely. We wish the phrase to apply rather broadly. In the spirit of this language then each of the lemmas below is a covering lemma. As we shall need to refer to these later on we present them as a series of lemmas. The proofs are elementary but, to familiarize the reader with the language, are included.

LEMMA 2.6 *Let β_1 and β_2 be full covering relations on a set E. Then $\beta_1 \cap \beta_2$ is a full covering relation on E.*

Proof. If β_1 and β_2 are full covering relations on a set E then there are, for every $x \in E$, positive numbers $\delta_1(x)$ and $\delta_2(x)$ so that $([y, z], x) \in \beta_1$ for every $y < x < z$ with $0 < z - y < \delta_1(x)$ and so that $([y, z], x) \in \beta_2$ for every $y < x < z$ with $0 < z - y < \delta_2(x)$. Evidently $([y, z], x) \in \beta_1 \cap \beta_2$ for every $y < x < z$ with $0 < z - y < \min\{\delta_1(x), \delta_2(x)\}$ and so $\beta_1 \cap \beta_2$ is full at each $x \in E$.

LEMMA 2.7 *Let β_1 be a full covering relation on a set E and let β_2 be a fine covering relation on E. Then $\beta_1 \cap \beta_2$ is a fine covering relation on E.*

Proof. Let us show that $\beta_1 \cap \beta_2$ is fine at any point $x \in E$. Since β_1 is a full covering relation on E then there is a positive number δ_1 so that $([y, z], x) \in \beta_1$ for every $y < x < z$ with $0 < z - y < \delta_1$. Now let $\epsilon > 0$ and set $\epsilon_1 = \min\{\delta_1, \epsilon\}$. Since β_2 is fine at x there is at least one pair $([y, z], x) \in \beta_2$

for which $y < x < z$ with $0 < z - y < \epsilon_1$. Evidently this same pair $([y, z], x)$ must belong also to β_1 and hence to the intersection $\beta_1 \cap \beta_2$. This shows that $\beta_1 \cap \beta_2$ is fine at each $x \in E$, as required.

LEMMA 2.8 *Let β be a full [fine] covering relation on a set E and let G be an open set containing E. Then $\beta(G)$ is a full [fine] covering relation on E.*

Proof.   Suppose that β is full at x and that $x \in E \subset G$ where G is open. Then there is a $\delta > 0$ so that $([y, z], x) \in \beta$ for every $y < x < z$ with $0 < z - y < \delta$. Let δ_1 be the smaller of δ and the distance from x to the complement of G. Then $([y, z], x) \in \beta(G)$ for every $y < x < z$ with $0 < z - y < \delta_1$; this follows from the fact that any such interval $[y, z]$ must be entirely contained in G. Hence $\beta(G)$ is full at x too and the lemma is proved under the assumption that β is full. The fine version is similar.

LEMMA 2.9 *Let β_α be a full [fine] covering relation on a set E_α for each $\alpha \in A$. Then*

$$\beta = \bigcup_{\alpha \in A} \beta_\alpha$$

is a full [fine] covering relation on the union of the $\{E_\alpha\}$.

Proof. If each β_α is a full covering relation on the set E_α then

$$\beta = \bigcup_{\alpha \in A} \beta_\alpha$$

must be full at every point in the union of the $\{E_\alpha\}$. For any x in that union there is an α with $x \in E_\alpha$; since β_α is full at such an x there is a positive number δ so that $([y, z], x) \in \beta_\alpha$ for every $y < x < z$ with $0 < z - y < \delta$. Of course then $([y, z], x) \in \beta$ for every $y < x < z$ with $0 < z - y < \delta$ and β is full at x as required. The fine version is similar.

2.3   Covering lemmas

We present now the fundamental covering lemmas that are characteristic of the basis we are investigating and which carry almost all of the deeper properties of the differentiation, measure and integration theory associated with this basis. These lemmas and the elementary ones from the preceding section provide all the technical properties of the differentiation basis needed to establish our results.

The first covering lemma may be loosely read as stating that if β is a full covering relation on a set E then one can write $E = \bigcup_1^\infty E_n$ in such a way that β is a *uniform* full covering relation on each E_n in the sense that there is a positive δ (depending on n) so that any pair (I, x) with $x \in I^0 \cap E_n$ and $|I| < \delta$ must belong to β. The importance of this technical looking fact lies in the way it allows local properties of functions on a set E to require certain uniform properties on subsets.

LEMMA 2.10 *Let β be a full covering relation on a set E. Then there is a denumerable partition $\{E_n\}$ of E so that $(I, x) \in \beta$ whenever I is an interval with $|I| < 1/n$, $x \in E_n$ and x is an interior point of I.*

Proof. Since β is a full covering relation on E there is a positive function δ defined on E so that whenever $|I| < \delta(x)$ and $x \in E$ is an interior point of I the pair (I, x) belongs to β. Let

$$E_n = \left\{ x \in E : n^{-1} \leq \delta(x) < (n-1)^{-1} \right\}.$$

If $x \in E_n$, $|I| < 1/n$ and x is an interior point of I then $|I| < \delta(x)$ so that $(I, x) \in \beta$.

Let us mention also the following variant of Lemma 2.10.

LEMMA 2.11 *Let β be a full covering relation on a set E. Then there is an increasing sequence of sets $\{E_n\}$ whose union includes the set E so that if $x \in E_n$, x is an interior point of an interval I and $|I| < 1/n$ then the pair (I, x) belongs to β.*

Recall that a finite packing π is a division of an interval $[a, b]$ if $\sigma(\pi) = [a, b]$ (see Section 2.1). Our next covering lemma asserts that full covering relations normally contain divisions of most intervals.

LEMMA 2.12 *Let β be a full covering relation on the real line. Then there is a countable set C so that β contains a division of every interval $[a, b]$ with $a, b \in \mathbb{R} \setminus C$.*

Proof. Let β be a full covering relation on an interval $[a, b]$. We shall show first that there is an $\epsilon > 0$ and a countable set $C \subset [a, b]$ so that β contains a division of all intervals $[c, d]$ with $a - \epsilon < c < a$ and $d \in (a, b) \setminus C$

Since β is a full covering relation on $[a, b]$ there is a positive number $\delta(x)$ for each $x \in [a, b]$ so that all pairs (I, x) with $x \in I^0$ and

$$I \subset (x - \delta(x), x + \delta(x))$$

belong to β.

Choose $\epsilon < \delta(a)$. Define T to be the set of points $z \in (a, b]$ so that β contains a division of $[c, w]$ for all $a - \epsilon < c < a$ and all but countably many w in (a, z). Clearly $T \supset (a, a + \epsilon)$; let $z' = \sup T$. By the nature of its definition $z' \in T$.

It cannot be that $z' < b$ for if so then, as we shall see, $T \supset (z', z' + \delta(z'))$ which contradicts $z' = \sup T$. For if $z' < b$ there is a division of $[c, c']$ for all $a - \epsilon < c < a$ and for some c' sufficiently close to z'; any pair $([c', c''], z')$ with $z' > c' > z' - \delta(z')$ and with $z' < c'' < z' + \delta(z')$ is in β. This shows that $z' = b$ so that the our assertion is proved.

Let β be a full covering relation on the line. From the above assertion we may now deduce that for every point $x \in \mathbb{R}$ there is a countable set C_x with the property that β contains a division of $[c, d]$ if $x \in (c, d)$, c is sufficiently close to x and $d \notin C_x$ or if d is sufficiently close to x and $c \notin C_x$. Let C denote the union of the sets C_r taken over all rational numbers r. Let $[x, y]$ denote any interval with endpoints not in C. We will show that β contains a division of every such interval $[x, y]$ and the lemma is proved.

Choose a rational $r \in (x, y)$ and choose $\delta_r > 0$ so that β contains a division of every interval $[x, z]$ for $r < z < r + \delta_r$. Now choose a rational s in this interval $(r, r + \delta_r)$ and argue in a similar way to obtain a $s - r > \delta_s > 0$ so that β contains a division of every interval $[z, y]$ for $s - \delta_s < z < s$. For such a z then β contains a division of both intervals $[x, z]$ and $[z, y]$; thus β contains a division of $[x, y]$ as required, and the proof is complete.

It is easy to check that the set C of the theorem may not be reduced to a finite set in general. Take $C = \{0, 1, 1/2, 1/3, \ldots\}$ and write

$$\beta = \{([a, b], x) : a < x < b, \ a, b \notin C, \ x \neq 0\} \cup \{([a, b], 0) : a < 0 < b\}.$$

Then β is full but contains no division of an interval with endpoints in C.

3   The variation

In real variable theory the notion of the variation of a function plays a key role. If h is an *interval function*, one that assigns a real value $h([a, b])$ to every nondegenerate interval $[a, b]$, then its total variation $H(I)$ on an interval $I = [c, d]$ is defined as the supremum of the sums

$$\sum_{i=1}^{n} |h([x_i, x_{i-1}])|$$

taken over all partitions $c = x_0 < x_1 < \ldots < x_n = d$. For the interval function $\Delta f : [a, b] \to f(b) - f(a)$ where f is a real function this is the familiar total variation for f from elementary real variables.

In the present setting this will be generalized to apply to any interval-point function (see Definition 3.1); it becomes especially useful when interpreted within the setting of covering relations. This variation is the fundamental concept on which the theory develops: it produces all the necessary measure theory and provides the link between derivation properties and measure-theoretic properties of functions.

The study of interval functions by now is well motivated. Many of the processes of analysis can be expressed with their help. Certainly the kind of derivation process we wish to investigate

$$\lim_{y,z \to x,\ y < x < z} \frac{|f(y) - f(z)|}{(y - z)^s},$$

can be directly viewed as a limit of an interval function (or as a limit of a quotient of interval functions). Many properties of such limits depend only on the form of the limit and not on the particular interval function. Integration procedures too have such an expression. Let f be a real function and define the interval functions h and H by $h([c, d]) = \inf_{x \in [c,d]} f(x)(d - c)$ and $H([c, d]) = \sup_{x \in [c,d]} f(x)(d - c)$. Then the upper and lower limits of the sums $\sum h([x_{i-1}, x_i])$ and $\sum H([x_{i-1}, x_i])$ taken over partitions of an interval $[a, b]$ express the upper and lower Darboux integrals of f.

However in some studies one needs to go somewhat beyond interval functions. For example the many variants on the classical Riemann integral study the limits of an expression

$$\sum_{i=1}^{n} f(\xi_i)(x_i - x_{i-1}).$$

Here there is some flexibility in using intervals $[x_{i-1}, x_i]$ and associated points $\xi_i \in [x_{i-1}, x_i]$ and the language that describes this process is that of covering relations. The expression $f(\xi_i)(x_i - x_{i-1})$ can be thought of conveniently as a function of pairs (I, x) which we shall call an interval-point function. Only functions of the form $(I, x) \to f(x)h(I)$ appear in applications but the theory is developed for general interval-point functions.

Our goal in this section is to develop standard properties of a pair of measures h^* and h_* associated with any interval-point function h. These measures will carry the variational information about h. We show that they are in general metric outer measures and so have useful topological properties. Since the notion of an interval-point function is very general we cannot expect regularity properties; for purely interval functions, however, we do obtain nicer properties.

3.1   Variation of an interval-point function

3.1 DEFINITION. An *interval-point function* is a real-valued function defined on some covering relation.

3.2 DEFINITION. Let β be a covering relation and let h be an interval-point function defined on β. Then by $\mathrm{Var}(h, \beta)$ we denote

$$\mathrm{Var}(h, \beta) = \sup \left\{ \sum_{(I, x) \in \pi} |h(I, x)| : \pi \subset \beta, \ \pi \text{ a packing} \right\}.$$

We refer to this as the *variation* of the function h taken relative to the covering relation β. Note that if β consists of all pairs (I, x) where I is any subinterval of an interval $[a, b]$ and h is an interval function, then $\mathrm{Var}(h, \beta)$ is exactly the variation of h on the interval $[a, b]$. Note too that the definition would be unchanged if π were restricted to be a finite packing. Normally in any computation we need consider only finite packings.

This variation is now refined by taking its limits over all full and fine covering relations; it is this refined variation that is the object of most of our study.

3.3 DEFINITION. Let h be an interval-point function and E a set of real numbers. Then we write

$$V^*(h, E) = \inf \left\{ \mathrm{Var}(h, \beta) : \beta \text{ a full covering relation on } E \right\}$$

and

$$V_*(h, E) = \inf \{\text{Var}(h, \beta) : \beta \text{ a fine covering relation on } E\}.$$

We have some additional suggestive notation for these two variations. We write h^* and h_* for the two set functions

$$h^*(E) = V^*(h, E)$$

and

$$h_*(E) = V_*(h, E)$$

and we refer to these set functions as the full and fine *variational measures* generated by h. It will be shown below that these are metric outer measures on the real line.

A restricted version of Definition 3.3 is also of some use.

3.4 DEFINITION. Let h be an interval-point function and X and E sets of real numbers. Then we write

$$V_X^*(h, E) = \inf \{\text{Var}(h, \beta(X)) : \beta \text{ a full covering relation on } E\}$$

and

$$V_{*X}(h, E) = \inf \{\text{Var}(h, \beta(X)) : \beta \text{ a fine covering relation on } E\}.$$

These are called the full and fine *restricted variation* of h.

Note that it is possible to write $V_X^*(h, E) = V^*(k, E)$ and $V_{*X}(h, E) = V_*(k, E)$ for a suitable choice of interval-point function k so that every property established for the variation applies immediately to the restricted variation.

3.2   Differential equivalence

Our study centers on the following equivalence relation. In the language of Henstock [7] this has been called "variational equivalence" and it is derives from the notion of "differential equivalence" of Kolmogorov [10]. It has been used more recently as a means to define a notion of "differential" (Leader [13] and [12]). Note that the equivalence relation depends on the differentiation basis being studied.

3.5 DEFINITION. Two interval-point functions h and k are said to be *equivalent* on a set E provided $V^*(h - k, E) = 0$. We write this symbolically as $h \equiv k$ on E.

If $V^*(h - k, R) = 0$ then we say merely that h and k are equivalent and write $h \equiv k$. That this is an equivalence relation follows from the fact that the variation is subadditive: if h_1 and h_2 are interval-point functions then

$$V^*(h_1 + h_2, E) \leq V^*(h_1, E) + V^*(h_2, E). \tag{5}$$

To prove the inequality (5) requires merely an application of the covering Lemma 2.6.

(The fine variation is not subadditive in general. For example take $h_1(I, x) = |I|$ if I has rational endpoints and $h_1(I, x) = 0$ otherwise. If $h_2(I, x) = |I| - h_1(I, x)$ then $V_*(h_1 + h_2, E) = |E|$ for any set E but $V_*(h_1, E) = V_*(h_2, E) = 0$.)

We conclude this section with an important relation. An interval function h is *additive* if $h(I \cup J) = h(I) + h(J)$ for any two abutting intervals I and J and *subadditive* if $h(I \cup J) \leq h(I) + h(J)$ for any such intervals I and J. It is *continuous* if at each point x and for each $\epsilon > 0$ there is a $\delta > 0$ so that $|h(I)| < \epsilon$ if $|I| < \delta$ and $x \in I^0$. (In Section 3.4 the notion of continuity will be extended to interval-point functions.)

THEOREM 3.6 *Let h be a continuous, nonnegative, subadditive interval function that has bounded variation in every compact interval and let $H(I)$ be the total variation of h on the interval I. If H too is continuous then $H \equiv h$.*

Proof. The total variation function H is finite valued, additive and $H \geq h$; we suppose too that H is continuous. We show that for any $\epsilon > 0$ there is a full covering relation β on the real line so that $\text{Var}(H - h, \beta) < \epsilon$. In each interval $[n, n + 1]$ we may place a finite number of points $x_{n,j}$ ($j = 0, 1, 2, \ldots, m_n$) in such a way that $x_{n,0} = n$, $x_{n,m_n} = n + 1$, and

$$\sum_{j=1}^{m_n} h([x_{n,j-1}, x_{n,j}]) \geq H([n, n + 1]) - \epsilon 2^{-|n|-2}.$$

Let C denote the (countable) collection of these points. Since H and h are continuous there is a full covering relation β_1 on C so that $\text{Var}(h, \beta_1) < \epsilon/4$

and $\mathrm{Var}(H, \beta_1) < \epsilon/4$. Let $G = R \setminus C$ and let β_2 be any full covering relation on the line. Then $\beta_1 \cup \beta_2(G)$ is a full covering relation on the line; if π is any packing in $\beta_1 \cup \beta_2(G)$ then set $\pi_1 = \pi[C]$ and $\pi_2 = \pi \setminus \pi_1$. We compute

$$\sum_{(I,x)\in\pi_1} H(I) - h(I) < \mathrm{Var}(h, \beta_1) + \mathrm{Var}(h, \beta_1) < \epsilon/2$$

and for each n

$$\sum_{(I,x)\in\pi_2([n,n+1])} H(I) - h(I) \leq H([n, n+1]) - \sum_{j=1}^{m_n} h([x_{n,j-1}, x_{n,j}])$$

so that

$$\sum_{(I,x)\in\pi_2} H(I) - h(I) \leq \sum_{n=-\infty}^{+\infty} \epsilon 2^{-|n|-2} = \epsilon/2.$$

This gives $V^*(H - h, R) \leq \mathrm{Var}(H - h, \beta) < \epsilon$ and, since ϵ is arbitrary, we must have $H \equiv h$ as required.

It should be noted that the assumption here that the total variation function H is continuous must be made in order for the conclusion $H \equiv h$ to hold. For example let $h([a, b]) = 0$ if $ab \neq 0$ and let $h([a, b]) = 1$ if $ab = 0$. Then h is continuous and subadditive. But the total variation of h is by direct computation the function H where $H([c, d]) = 0$ if $cd > 0$, $H([c, 0]) = H([0, d]) = 1$ and $H([c, d]) = 2$ if $c < 0 < d$. Clearly H is not continuous and $H \not\equiv h$.

3.3   Variational measures

The full and fine variations h^* and h_* associated with any interval-point function h are genuine outer measures on the real line that have nice topological properties.

THEOREM 3.7 *For any interval-point function h the set functions h^* and h_* are metric outer measures.*

Proof. We show first that the set function h^* is countably subadditive. Let $X, Y_1, Y_2, Y_3, \ldots$ be a sequence of sets for which $X \subset \bigcup_{i=1}^{\infty} Y_i$. Suppose that a positive number ϵ has been given and choose β_i full covering relations on the sets Y_i in such a way that

$$\mathrm{Var}(h, \beta_i) \le h^*(Y_i) + \epsilon/(2^i). \tag{6}$$

Define β as the union of the families β_i then, by Lemma 2.9, β is a full covering relation on the set X. Therefore, using the inequalities in (6), we must have

$$h^*(X) \le \mathrm{Var}(h, \beta) \le \sum_{i=1}^{\infty} \mathrm{Var}(h, \beta_i) \le \sum_{i=1}^{\infty} h^*(Y_i) + \epsilon.$$

As ϵ is an arbitrary positive number we have then the inequality $h^*(X) \le \sum_{i=1}^{\infty} h^*(Y_i)$ as required to establish that h^* is an outer measure. In precisely the same manner it may be shown that h_* is also an outer measure.

To see that h^* is a metric outer measure (see [15, Definition 14, p. 28]) suppose that two sets X_1 and X_2 are separated in such a way that there are open sets G_1 and G_2 with $X_1 \subset G_1$, $X_2 \subset G_2$, and $G_1 \cap G_2 = \emptyset$. Choose a full covering relation β on the set $X_1 \cup X_2$ such that

$$Var(h, \beta) \le h^*(X_1 \cup X_2) + \epsilon.$$

Then the collections $\beta_1 = \beta(G_1)$ and $\beta_2 = \beta(G_2)$ are full covering relations on X_1 and X_2 because of Lemma 2.8. Now we compute

$$h^*(X_1) + h^*(X_2) \le \mathrm{Var}(h, \beta_1) + \mathrm{Var}(h, \beta_2) \le \mathrm{Var}(h, \beta)$$

since the covering relations β_1 and β_2 are "separated", i.e. if $(I, x) \in \beta_1$ and $(J, y) \in \beta_2$ the intervals I and J are disjoint. Hence $h^*(X_1) + h^*(X_2) \le h^*(X_1 \cup X_2) + \epsilon$ and since $\epsilon > 0$ is arbitrary we must have the inequality

$$h^*(X_1) + h^*(X_2) \le h^*(X_1 \cup X_2).$$

As the opposite inequality is immediate an equality must hold as required to show that h^* is a metric outer measure. The same arguments apply to h_* .

THEOREM 3.8 *Let B be a Borel set with $h^*(B) < +\infty$. Then for every $\epsilon > 0$ there is a closed set C contained in B for which*

$$h^*(B \setminus C) < \epsilon.$$

The same assertion holds for the measure h_*.

Proof. This is a standard measure theoretic result available for any metric outer measure; see [5, 2.2.2, p.60].

THEOREM 3.9 *If B is a Borel set contained in a countable union of open sets of finite h^*–measure then there is an open set W containing B so that*

$$h^*(W \setminus B) < \epsilon.$$

The same assertion holds for the measure h_*.

Proof. This too is a standard measure theoretic result (*op.cit.*).

3.4   Increasing sets property

In this section we prove that the outer measures h^* and h_* have the increasing sets property.

We require first a notion of continuity for interval-point functions. Let us say that h is *continuous* at a point x_0 provided that $h^*(\{x_0\}) = 0$ and that h is *weakly continuous* at a point x_0 provided that $h_*(\{x_0\}) = 0$. This can be written directly in terms of the interval-point function h by noticing that

$$\limsup_{I \to x} |h(I, x_0)| = h^*(\{x_0\})$$

and

$$\liminf_{I \to x} |h(I, x_0)| = h_*(\{x_0\})$$

where the limit is taken over all sequences of intervals $\{I_k\}$ with $|I_k| \to 0$ and x_0 an interior point of each I_k. This notion clearly extends the definition of continuity given for interval functions in Section 3.2.

THEOREM 3.10 *Let h be a continuous interval-point function and E_i an increasing sequence of sets. Then*

$$h^* \left(\bigcup_{i=1}^{\infty} E_i \right) = \lim_{n \to +\infty} h^*(E_n).$$

This follows from a series of lemmas which we now prove. We need a type of "uniform" property of the restricted variation. Our first lemma shows how the variation splits at sets K of a given form, and the second gives the uniformity result.

LEMMA 3.11 *Let h be a continuous interval-point function, E a set of real numbers, K a finite union of closed intervals and H the complement of the interior of K. Then*

$$V^*(h, E) = V_K^*(h, E) + V_H^*(h, E).$$

Proof. Let C denote the (finite) set of endpoints of K and let $\epsilon > 0$. Since h is continuous we may choose a full covering relation β_1 on C so that $\text{Var}(h, \beta_1) < \epsilon$. Let β_2 and β_3 be arbitrary full covering relations on E. Define

$$\beta_4 = \beta_1 \cup \beta_2(K) \cup \beta_3(H).$$

Note that β_4 is a full covering relation on E and so

$$V^*(h, E) \leq \text{Var}(h, \beta_4) \leq \text{Var}(h, \beta_1) + \text{Var}(h, \beta_2(K)) + \text{Var}(h, \beta_3(H))$$

from which we deduce

$$V^*(h, E) \leq \text{Var}(h, \beta_2(K)) + \text{Var}(h, \beta_3(H)) + \epsilon.$$

From this the assertion of the lemma may be obtained.

Note that we may also prove the following variant of Lemma 3.11. Let J_1 and J_2 be nonoverlapping intervals ; then for any continuous interval-point function h

$$V_{J_1 \cup J_2}^*(h, E) = V_{J_1}^*(h, E) + V_{J_2}^*(h, E).$$

LEMMA 3.12 *Let $\epsilon > 0$ and let h be a continuous interval-point function. Suppose that $V^*(h, E) < +\infty$ and that β is a full covering relation on E. If the inequality*

$$Var(h, \beta) \leq V^*(h, E) + \epsilon$$

holds, then for any set K that is a finite union of closed intervals

$$Var(h, \beta(K)) \leq V_K^*(h, E) + \epsilon.$$

Proof. If we have the inequality $\text{Var}(h, \beta) \leq V^*(h, E) + \epsilon$, if K is a finite union of intervals and if H is the complement of the interior of K then, by Lemma 3.11,

$$
\begin{aligned}
\text{Var}(h, \beta(K)) &\leq \text{Var}(h, \beta) - \text{Var}(h, \beta(H)) \\
&\leq V^*(h, E) + \epsilon - V_H^*(h, E) \\
&= V_K^*(h, E) + \epsilon
\end{aligned}
$$

as required.

We now prove the statement of Theorem 3.10. Let us suppose that $\{E_n\}$ is an increasing sequence of sets and we may assume that each $h^*(E_n)$ is finite. Let $\epsilon > 0$ and choose for each index n a full covering relation β_n on E_n in such a way that

$$
\text{Var}(h, \beta_n) \leq h^*(E_n) + \epsilon/2^n.
$$

Construct the family β as

$$
\beta = \beta_1[E_1] \cup \beta_2[E_2 \setminus E_1] \cup \beta_3[E_3 \setminus E_2] \cdots
$$

and note that, by Lemma 2.9, β is a full covering relation on the set $E = \bigcup_{n=1}^{\infty} E_n$. Let $\pi \subset \beta$ be any finite packing. We write $\pi_k = \pi[E_k \setminus E_{k-1}]$ and let N be the first index for which $\pi_k = \emptyset$ for $k > N$. Let K_k be the union of the intervals in π_k. Now we compute, using the uniformity result from Lemma 3.12 above,

$$
\begin{aligned}
\sum_{i=1}^{m} |h(I_i, x_i)| &= \sum_{k=1}^{N} \sum_{(I,x) \in \pi_k} |h(I, x)| \\
&\leq \sum_{k=1}^{N} \text{Var}(h, \beta_k) \\
&\leq \sum_{k=1}^{N} \left\{ V_{K_k}^*(h, E_k) + \epsilon/2^k \right\} \\
&\leq \sum_{k=1}^{N} V_{K_k}^*(h, E_N) + \epsilon \\
&\leq V^*(h, E_N) + \epsilon \\
&\leq \sup \{ h^*(E_n) : n = 1, 2, 3, \ldots \} + \epsilon.
\end{aligned}
$$

This estimate is available for any such choice of π from β and $\epsilon > 0$ is arbitrary. Hence we obtain that $h^*(E) \leq \sup_n h^*(E_n)$ as required to prove Theorem 3.10.

The property of Theorem 3.10 is also available for the measure h_* with a similar hypothesis. The proof is almost identical to that given above replacing full covering relations by fine covering relations.

THEOREM 3.13 *Let h be a weakly continuous interval-point function and E_i an increasing sequence of sets. Then*

$$h_*(\bigcup_{i=1}^{\infty} E_i) = \lim_{n \to +\infty} h_*(E_i).$$

3.5 Regularity properties

The increasing sets property expressed in Theorem 3.10 is possessed by any regular outer measure. Recall that an outer measure μ is said to be *regular* if for every set X there is a μ–measurable set Y containing X with $\mu(X) = \mu(Y)$. It is said to be regular with respect to some class of measurable sets (eg. the Borel sets, the $\mathcal{G}_\delta$ sets, the $\mathcal{G}_{\delta\sigma}$ sets) if the same equality can be obtained within the smaller class of sets. It is easy to extend certain properties that μ will then have on that smaller class to hold on all sets. Usually regularity is not too difficult to obtain and the increasing sets property is an immediate corollary. Here regularity questions are rather more delicate.

Let us show that neither of the outer measures h^* and h_* is in general Borel regular, but that the former outer measure is so if h is an interval function.

THEOREM 3.14 *In general h^* and h_* need not be Borel regular.*

Proof.   Let E be a non Borel set and define $h(I, x)$ to be 0 if x is in E and to be 1 otherwise. Then there is a full covering relation β_1 on E for which $\text{Var}(h, \beta_1) = 0$. This gives that $h^*(E) = 0$. But if E is a proper subset of any set F and β_2 is a full covering relation on F then there must be in β_2 at least one pair (I, x) with $h(I, x) = 1$ and hence $\text{Var}(h, \beta) \geq 1$. Thus $h^*(F) \geq 1 > h^*(E)$. Thus there can be no Borel set that contains E and has the same h^* –measure. The same assertion is available for the measure h_*.

THEOREM 3.15 *Suppose that h is an interval function. Then the measure h^* is $\mathcal{F}_{\sigma\delta}$ regular.*

Proof. Let E be a set of real numbers with $h^*(E) < \infty$. For any $\epsilon > 0$ we may choose a full covering relation β on E so that

$$\mathrm{Var}(h, \beta) \leq h^*(E) + \epsilon.$$

We use Lemma 2.11 to obtain an increasing sequence of sets E_n whose union is E such that if the interval I has at least one interior point in E_n and $|I| < 1/n$ then there is a pair (I, x) in β .

The collection γ_n of all pairs (I, z) for $|I| < 1/n$, $z \in \overline{E}_n \cap I^0$ is a (uniform) full covering relation on the set $\overline{E}_n$ and hence we may conclude, remembering that $h(I, x)$ depends only on I,

$$h^*(\overline{E}_n) \leq \mathrm{Var}(h, \gamma_n) \leq \mathrm{Var}(h, \beta) \leq h^*(E) + \epsilon.$$

and from this we see that

$$h^*(\bigcup_1^\infty \overline{E}_n) = \lim_{n \to \infty} h^*(\overline{E}_n) \leq h^*(E) + \epsilon.$$

From this inequality we see that the measure $h^*(E)$ may be obtained from a larger set that is an intersection of a countable union of closed sets as asserted in the statement of the theorem.

THEOREM 3.16 *Suppose that h is a nonnegative, additive interval function. Then the measure h^* is $\mathcal{G}_\delta$ regular.*

Proof. Note first that for any interval (a, b),

$$h^*((a, b)) \leq \lim_{c \to a+, d \to b-} h([c, d]). \tag{7}$$

Suppose that $h^*(E)$ is finite and let $\epsilon > 0$. Let β be a full covering relation on E chosen so that $\mathrm{Var}(h, \beta) < h^*(E) + \epsilon$. Let the positive function δ be defined so that whenever $|I| < 2\delta(x)$ and $x \in I^0 \cap E$ then $(I, x) \in \beta$. Define the open set

$$G = \bigcup_{x \in E} (x - \delta(x), x + \delta(x)).$$

Suppose that (a, b) is a component interval of G and that $[c, d]$ is a compact subinterval. Then there is a finite collection

$$\{(x_i - \delta(x_i), x_i + \delta(x_i)) : i = 1, 2, \ldots n\}$$

that covers $[c, d]$. We may make this minimal so that no interval contains another and label the points so that $x_1 < x_2 \ldots < x_n$; then a simple geometric argument shows that there is a packing $\pi = \{(I_i, x_i)\}$ with $x_i \in I_i^0$, $|I_i| < 2\delta(x_i)$ and $\sigma(\pi) \supset [c, d]$. Such a packing belongs to β and this shows that $h([c, d])) \leq \mathrm{Var}(h, \beta((a, b)))$. for all $a < c < d < b$. Combining this with (7) we obtain $h^*((a, b)) \leq \mathrm{Var}(h, \beta((a, b)))$. Now summing over all the component intervals of G we obtain $h^*(G) \leq \mathrm{Var}(h, \beta) \leq h^*(E) + \epsilon$ from which we may conclude that the measure h^* is $\mathcal{G}_\delta$ regular.

3.6   The upper integral

The functional

$$f \to V^*(fh, E)$$

for a fixed interval-point function h and a fixed set E behaves as an upper integral defined for all nonnegative real functions f. Let us introduce the following convenient notation for this concept and develop its properties: we write

$$h^*(f) = V^*(fh, \mathbb{R})$$

and

$$h_*(f) = V_*(fh, \mathbb{R})$$

and we refer to these as the *full* and *fine upper integrals* associated with the interval-point function h. Here f is an arbitrary nonnegative real function. Note this does not interfere with the notation used in the preceding sections for the variational measure because of the relations $h^*(E) = h^*(\chi_E)$ and $h_*(E) = h_*(\chi_E)$ for any set $E \subset R$.

We study the upper integral h^* for an arbitrary continuous interval-point function h; continuity can be avoided but is the most natural assumption for this particular differentiation basis. The basic properties of this upper integral are developed in this section. Assertions 1 – 4 of the first theorem express properties that in the literature are usually assumed for "upper integrals" (cf. [24]) and accordingly we have chosen this nomenclature.

THEOREM 3.17 *For any continuous interval-point function h and any nonnegative real functions $f, g_1, g_2, \ldots$ the following assertions are true:*

1. $0 \le h^*(f) \le +\infty$.

2. *If* $0 \le f_1(x) \le f_2(x)$ *at each point x then* $h^*(f_1) \le h^*(f_2)$.

3. $h^*(cf) = c(h^*(f))$ *for any $c > 0$* .

4. $h^*(f_1 + f_2) \le h^*(f_1) + h^*(f_2)$.

5. *If* $g_1(x) \le g_2(x) \le g_3(x) \le \ldots$ *and* $f(x) \le \sup_n g_n(x)$ *hold at each point x then*
$$h^*(f) \le \lim_{n \to +\infty} h^*(g_n).$$

Proof. The first three of these are elementary and the fourth was established in Section 3.2. In order to prove the fifth we may suppose that $h^*(g_n)$ is finite for each n. Let $\epsilon > 0$ and $0 < c < 1$. For each point x there is a least integer $n(x)$ for which $cf(x) \le g_m(x)$ if $m \ge n(x)$. Choose a sequence $\{\beta_n\}$ so that each is a full covering relation on $\mathbb{R}$ with the property that

$$\mathrm{Var}(g_n h, \beta) \le V^*(g_n h, \mathbb{R}) + \epsilon/2^n.$$

By Lemma 3.12 this gives

$$\mathrm{Var}(g_n h, \beta(K)) \le V_K{}^*(g_n h, \mathbb{R}) \qquad (8)$$

for every finite union of intervals K. Define the sets $X_n = \{x : n(x) = n\}$ and define the collection $\beta = \bigcup_{n=1}^\infty \beta_n[X_n]$. Since the sequence of sets $X_1, X_2, \ldots$ is disjoint and covers the entire real line, β must be a full covering relation on R. We now estimate $V^*(fh, R)$ by computing $\mathrm{Var}(fh, \beta)$. Let π denote an arbitrary finite packing contained in β and write $K_n = \sigma(\pi_n)$ where $\pi_n = \pi[X_n]$. There is a first integer N so that π_m is empty for all $n > N$.

Now we compute, using (8),

$$\sum_{(I,x)\in\pi} |f(x)h(I,x)| = \sum_{i=1}^N \sum_{(I,x)\in\pi_n} |f(x)h(I,x)|$$

$$\le c^{-1} \sum_{i=1}^N \sum_{(I,x)\in\pi_n} |g_n(x)h(I,x)|$$

$$\leq \quad c^{-1} \sum_{i=1}^{N} \text{Var}(g_n h, \beta_n(K_n))$$

$$\leq \quad \epsilon c^{-1} + c^{-1} \left\{ \sum_{i=1}^{N} V_{K_n}{}^*(g_n h, \mathbb{R}) \right\}$$

$$\leq \quad \epsilon c^{-1} + c^{-1} \left\{ \sum_{i=1}^{N} V_{K_n}{}^*(g_N h, \mathbb{R}) \right\}$$

$$\leq \quad \epsilon c^{-1} + c^{-1} \left\{ V^*(g_N h, \mathbb{R}) \right\}.$$

As this holds for all finite packings $\pi \subset \beta$ we have

$$V^*(fh, \mathbb{R}) \leq \text{Var}(fh, \beta) \leq \epsilon c^{-1} + c^{-1} \left\{ V^*(g_N h, \mathbb{R}) \right\},$$

and hence that

$$V^*(fh, \mathbb{R}) \leq \epsilon c^{-1} + c^{-1} \left\{ \sup_N V^*(g_N h, \mathbb{R}) \right\}.$$

Letting $\epsilon \to 0+$ and $c \to 1-$ in this inequality we obtain the result.

THEOREM 3.18 *For any continuous interval-point function h and any nonnegative real function f the upper integral $h^*(f)$ vanishes if and only if $f(x) = 0$ for h^*–almost every $x \in \mathbb{R}$.*

Proof. Let g_n denote the real function defined by $g_n(x) = f(x)$ if $f(x) < n$ and otherwise $g_n(x) = 0$. Then if $h^*(X) = 0$ we have

$$(fh)^*(X) = \lim_{n \to \infty} (g_n h)^*(X)$$

and

$$(g_n h)^*(X) \leq n h^*(X) = 0$$

so that $(fh)^*(X) = 0$. Thus if $f(x) = 0$ for all x in a a set Z with $h^*(\mathbb{R} \backslash Z) = 0$ then $(fh)^*(\mathbb{R}) \leq (fh)^*(Z) + (fh)^*(\mathbb{R} \backslash Z)$ and $(fh)^*(Z) = 0$ give that $h^*(f) = 0$ as required.

Conversely if $h^*(f) = 0$ then define the sets

$$X_n = \{x : f(x) > 1/n\}.$$

Since $h^*(X_n) \leq V^*(nfh, X_n) \leq n h^*(f) = 0$ the set of points x at which $f(x) \neq 0$ is the union of a sequence of sets having h^*–measure zero.

THEOREM 3.19 *Let h be a continuous interval-point function and let $f, g_1, g_2, \ldots$ be a sequence of functions such that $0 \leq f(x) \leq \sum_{n=1}^{\infty} g_n(x)$ everywhere. Then*

$$h^*(f) \leq \sum_{n=1}^{\infty} h^*(g_n).$$

Proof. This is evidently true for finite sums and the extension to infinite sums requires merely an application of Theorem 3.17.

These results extend to the dual upper integral h_* under weak continuity assumptions. Mostly the same proofs apply. While most of the properties extend to this dual functional, the subadditivity result,

$$h_*(f_1 + f_2) \leq h_*(f_1) + h_*(f_2)$$

would not be valid. A partial subadditivity result is available from the measure-theoretic representation given in the next section should these functions be h_*–measurable. The remaining results can be proved. In particular the analogues of Theorems 3.18 and 3.17 are obtained by essentially the same arguments. We state these without proofs.

THEOREM 3.20 *Let h be an interval-point function that is weakly continuous everywhere and let $f, g_1, g_2, \ldots$ be a sequence of functions with $g_1(x) \leq g_2(x) \leq g_3(x) \leq \ldots$ and $f(x) \leq \sup_n g_n(x)$ holding everywhere. Then*

$$h_*(f) \leq \lim_{n \to +\infty} h_*(g_n).$$

THEOREM 3.21 *Let h be weakly continuous everywhere. Then $h_*(f) = 0$ if and only if $f(x) = 0$ for h_*–almost every point x.*

3.7 The variational measure as an integral

Given an interval-point function h and a nonnegative real function f one might construct the outer measure $(fh)^*$ and consider that in some sense this is the "product" of the measure h^* and the real function f. If f is h^*–measurable then an alternative and more familiar construction, from a measure-theoretic viewpoint, would be to take as a form of product the measure

$$\nu(E) = (L) \int_E f(x) \, dh^*(x)$$

defined for all h^*–measurable sets E. Here the integral is in the usual Lebesgue sense and thus requires the h^*–measurability of the set E and of the function f. These two forms of product are closely related. Our main result in this section connects our variational computations with the standard measure-theoretic computations that are available when a measure has been constructed. The dual version using the measure h_* is given in the subsequent section.

THEOREM 3.22 *Let h be a continuous interval-point function, let E be a h^*-measurable set and let f be a nonnegative h^*-measurable real function. Then*

$$(fh)^*(E) = h^*(f\chi_E) = (L)\int_E f(x)\, dh^*(x).$$

Proof. If E_1 and E_2 are disjoint sets of real numbers with the set E_1 being h^*–measurable and c_1 and c_2 are any positive real numbers then we shall show that

$$h^*(c_1\chi_{E_1} + c_2\chi_{E_2}) = c_1 h^*(E_1) + c_2 h^*(E_2). \tag{9}$$

To prove this let $\epsilon > 0$ be given and choose a full covering relation β on $E_1 \cup E_2$ in such a way that

$$h^*(E_1) + \epsilon/2 \geq \text{Var}(h, \beta[E_1]),$$
$$h^*(E_2) + \epsilon/2 \geq \text{Var}(h, \beta[E_2]),$$

and

$$h^*(c_1\chi_{E_1} + c_2\chi_{E_2}) + \epsilon \geq \text{Var}(h(c_1\chi_{E_1} + c_2\chi_{E_2}), \beta).$$

There must be a packing $\pi \subset \beta$ so that

$$\sum_{(I,x)\in\pi} |h(I,x)| \geq h^*(E_1 \cup E_2) - \epsilon/2 = h^*(E_1) + h^*(E_2) - \epsilon/2.$$

If we write $\pi_1 = \pi[E_1]$ and $\pi_2 = \pi[E_2]$ then

$$\sum_{(I,x)\in\pi_1} |h(I,x)|$$
$$= \sum_{(I,x)\in\pi} |h(I,x)| - \sum_{(I,x)\in\pi_2} |h(I,x)|$$
$$\geq h^*(E_1) + h^*(E_2) - \epsilon/2 - [h^*(E_2) + \epsilon/2]$$
$$= h^*(E_1) - \epsilon.$$

Thus we have the inequality

$$\sum_{(I,x)\in\pi_1} |h(I,x)| \geq h^*(E_1) - \epsilon$$

and so similarly

$$\sum_{(I,x)\in\pi_2} |h(I,x)| \geq h^*(E_2) - \epsilon.$$

From these inequalities and the preceding estimates we have

$$
\begin{aligned}
&h^*(c_1\chi_{E_1} + c_2\chi_{E_2}) + \epsilon \\
&\geq \operatorname{Var}(h(c_1\chi_{E_1} + c_2\chi_{E_2}), \beta) \\
&\geq \sum_{(I,x)\in\pi} |c_1\chi_{E_1}(x)h(I,x) + c_2\chi_{E_2}(x)h(I,x)| \\
&= \sum_{(I,x)\in\pi_1} |c_1 h(I,x)| + \sum_{(I,x)\in\pi_2} |c_2 h(I,x)| \\
&\geq c_1 h^*(E_1) + c_2 h^*(E_2) - (c_1 + c_2)\epsilon.
\end{aligned}
$$

Since ϵ is an arbitrary positive number and since the inequality goes easily the other way we have established equation (9).

Now suppose that $E_1, E_2, E_3, \ldots$ is a disjoint sequence of h^*-measurable sets and that $c_1, c_2, c_3, \ldots$ a sequence of positive numbers. Write

$$g_n(x) = \sum_{i=1}^{n} c_i \chi_{E_i}(x).$$

Arguments similar to those given for equation (9) allow us to conclude that

$$h^*(g_n) = \sum_{i=1}^{n} c_i h^*(E_i). \tag{10}$$

Now if g is a nonnegative h^*–elementary function given in the form

$$g(x) = \sum_{i=1}^{\infty} c_i \chi_{E_i}(x)$$

for a disjoint sequence of h^*–measurable sets E_i then Theorem 3.17 and equation (10) show that

$$h^*(g) = \sum_{i=1}^{\infty} c_i h^*(E_i) = \int_R g(x)\, dh^*(x). \tag{11}$$

Now let f be an arbitrary bounded, nonnegative h^*-measurable function. Certainly, because of equation (11), we can obtain the inequality

$$h^*(f) \leq \int_R f(x) \, dh^*(x). \tag{12}$$

To obtain equality in inequality (12) we suppose that $f(x) < M$ everywhere and that $r > 1$. Define the sets

$$E_n = \left\{ x : Mr^{-n} \leq f(x) < Mr^{-n+1} \right\}$$

for $n = 1, 2, 3, \ldots$. This is certainly a disjoint sequence of h^*-measurable sets and the function g defined by setting

$$g(x) = \sum_{i=1}^{\infty} Mr^{-n+1} \chi_{E_n}(x)$$

is h^*-elementary and satisfies $f \leq g \leq rf$. Thus we have

$$h^*(f) \leq h^*(g) = \int_R g(x) \, dh^*(x) \leq rh^*(f).$$

From the definition of the integral we obtain then

$$h^*(f) = \int_R f(x) \, dh^*(x)$$

for all bounded, nonnegative h^*-measurable functions f. The extension to unbounded functions follows the usual measure-theoretic device by using the monotone convergence result in Theorem 3.17 for the upper integral.

3.8   The fine variational measure as an integral

We now obtain a version of Theorem 3.22 for the fine variational measure h_* under additional measurability assumptions and weak continuity assumptions on h.

THEOREM 3.23 *Let h be a weakly continuous interval-point function, E a Borel set and f a nonnegative Borel function. Then*

$$(fh)_*(E) = h_*(f\chi_E) = \int_E f(x) \, dh_*(x).$$

Proof. The proof is somewhat similar to the proof of Theorem 3.22 but we must replace the assertion (9) by a similar assertion; suppose that E_1 and E_2 are disjoint bounded Borel sets and c_1 and c_2 are positive real numbers. Then

$$h_*(c_1\chi_{E_1} + c_2\chi_{E_2}) = c_1 h_*(E_1) + c_2 h_*(E_2). \tag{13}$$

To prove this in one direction is easy. If fact if f_1 and f_2 are any nonnegative functions with $f_1(x)f_2(x) = 0$ everywhere then

$$h_*(f_1 + f_2) \leq h_*(f_1) + h_*(f_2). \tag{14}$$

Let A_1 and A_2 be the sets of points at which f_1 and f_2 respectively do not vanish and let β_1 and β_2 be any fine covering relations on A_1 and A_2. Then

$$
\begin{aligned}
h_*(f_1 + f_2) &\leq \operatorname{Var}((f_1 + f_2)h, \beta_1[A_1] \cup \beta_2[A_2]) \\
&\leq \operatorname{Var}((f_1 + f_2)h, \beta_1[A_1]) + \operatorname{Var}((f_1 + f_2)h, \beta_2[A_2]) \\
&= \operatorname{Var}(f_1 h, \beta_1[A_1]) + \operatorname{Var}(f_2 h, \beta_2[A_2])
\end{aligned}
$$

and (14) follows.

Thus, in order to establish (13), we shall require only the inequality in the other direction. We may suppose that these sets have finite measure. Let $\epsilon > 0$ be given; by Theorem 3.8 we may choose compact sets C_1 and C_2 contained in E_1 and E_2 respectively so that $h_*(E_1 \setminus C_1) < \epsilon$ and $h_*(E_2 \setminus C_2) < \epsilon$. Let β be an arbitrary fine covering relation on $E_1 \cup E_2$. If G_1 and G_2 are disjoint open sets containing C_1 and C_2 respectively then $\beta(G_1)$ is a fine covering relation on C_1 and $\beta(G_2)$ is a fine covering relation on C_2. Now we compute

$$
\begin{aligned}
c_1 h_*(E_1) &+ c_2 h_*(E_2) - \epsilon(c_1 + c_2) \\
&\leq c_1 h_*(C_1) + c_2 h_*(C_2) \\
&\leq \operatorname{Var}(c_1 h\chi_{E_1}, \beta(G_1)) + \operatorname{Var}(c_2 h\chi_{E_2}, \beta(G_2)) \\
&= \operatorname{Var}(c_1 h\chi_{E_1} + c_2 h\chi_{E_2}, \beta(G_1 \cup G_2)) \\
&\leq \operatorname{Var}(c_1 h\chi_{E_1} + c_2 h\chi_{E_2}, \beta).
\end{aligned}
$$

Since β is an arbitrary fine covering relation the equation (13) will follow.

The remaining parts of the proof of the theorem are duplications of the arguments used in the proof of Theorem 3.22.

3.9   Differential equivalence

Recall that the equivalence relation "$h \equiv k$ on E" means that $V^*(h-k, E) = 0$. There are some immediate properties that we may obtain by translating some earlier results. Throughout h_1 and h_2 are interval-point functions and f is a real function.

Certainly if $h_1 \equiv h_2$ on a set E then $h_1 \equiv h_2$ on any subset of E. The equivalence relation is additive over sequences of sets: if $h_1 \equiv h_2$ on each set E_n for $n = 1, 2, \ldots$ then $h_1 \equiv h_2$ on the set $E = \bigcup_{n \geq 1} E_n$. This follows from elementary properties of the variation. The relation $h_1 \equiv h_2$ may be multiplied by an arbitrary function: if $h_1 \equiv h_2$ on a set E then for any real function f the equivalence $fh_1 \equiv fh_2$ holds on E. The proof of this is contained in the more general statement below which we present as a lemma.

LEMMA 3.24 *If $h_1 \equiv h_2$ on a set E and k is an interval-point function for which*

$$\limsup_{I \to x} |k(I, x)| < +\infty$$

at every point $x \in E$ then $kh_1 \equiv kh_2$ on E.

Proof. Let X_n denote the set of points $x \in E$ at which

$$\limsup_{I \to x} |k(I, x)| < n$$

and define $\beta_n = \{(I, x) : |k(I, x)| < n\}$. Because of the assumptions in the statement of the theorem this is a full covering relation on X_n; for any other full covering relation β on E we have

$$\mathrm{Var}(kh_1 - kh_2, \beta \cap \beta_n[X_n]) \leq n\,\mathrm{Var}(h_1 - h_2, \beta).$$

From this we obtain $V^*(kh_1 - kh_2, X_n) = 0$ for all integers n and since the sets X_n evidently cover E the conclusion of the theorem will follow.

There are other useful observations on this equivalence relation. For any interval-point function h and any real function f the relation $fh \equiv 0$ on E holds if and only if $f(x) = 0$ for h^*–almost every point $x \in E$. More generally the relation $fh \equiv gh$ on E holds if and only if $f(x) = g(x)$ for h^*–almost every point $x \in E$. For an additive interval function h the relation $h \equiv 0$ can hold only if $h([a, b]) = 0$ for all pairs a, b not in some countable set. This last comment follows directly from Lemma 2.12.

Perhaps more important from our viewpoint is the fact that the equivalence relation preserves the measures.

THEOREM 3.25 *For any interval-point functions h and k if $h \equiv k$ on a set E then the full and fine variational measures for h and k are identical: $h^*(E) = k^*(E)$ and $h_*(E) = k_*(E)$.*

Proof. If $h \equiv k$ on E then $V^*(h - k, E) = 0$. Let $\epsilon > 0$ and choose a full covering relation β on E so that $\mathrm{Var}(h - k, \beta) < \epsilon$. Then if β_1 is any full (fine) covering relation on E the collection $\beta_1 \cap \beta$ is also a full (fine) covering relation on E and

$$\begin{aligned} \mathrm{Var}(h, \beta \cap \beta_1) &\le \mathrm{Var}(k, \beta \cap \beta_1) + \mathrm{Var}(h - k, \beta) \\ &< \mathrm{Var}(k, \beta_1) + \epsilon. \end{aligned}$$

This shows that $h^*(E) \le k^*(E)$ and so by symmetry equality must hold. A similar proof may be constructed for the dual measures.

3.10   A density theorem

The density theorem we now obtain for the h_* measure is modeled after the well known Lebesgue density theorem for Lebesgue measure. If h is the interval function $[a, b] \to b - a$ then h_* will be seen (in Section 6.1) to be exactly Lebesgue outer measure and so the theorem will reduce to the Lebesgue density theorem in this special case. It will include as well standard density theorems for the Hausdorff measures: if h is the interval function $[a, b] \to (b - a)^s$ then h_* will be seen (in Section 6.3) to be directly related to the Hausdorff s–dimensional measure.

THEOREM 3.26 *Let h be an interval function and let E be a Borel set of finite h_*–measure. Then*

$$\limsup_{I \to x} \frac{h_*(E \cap I)}{|h(I)|} = 0$$

for h_*-almost every point $x \notin E$ and

$$\limsup_{I \to x} \frac{h_*(E \cap I)}{|h(I)|} = 1$$

for h_*-almost every point $x \in E$.

Proof. Throughout the proof we do not dwell on proving the measurability of the sets that arise; the material in Section 4.2 can be used to show that the sets are even Borel. If the first assertion of the theorem is not true then for some sufficiently small positive number ϵ the set E_1 of points x not in E for which

$$\limsup_{I \to x} \frac{h_*(E \cap I)}{|h(I)|} > \epsilon$$

has h_*-measure greater than ϵ. By Theorem 3.8 there is a closed set $E_2 \subset E$ so that $h_*(E \setminus E_2) < \epsilon^2$.

The collection

$$\beta = \{(I, x) : h_*(E \cap I) \geq \epsilon|h(I)|\}$$

is a fine covering relation on E_1 and so also must be $\beta(G)$ where G is the complement of E_2. Thus

$$h_*(E_1) \leq \mathrm{Var}(h, \beta(G)) \leq \epsilon^{-1} h_*(E \cap G) \leq \epsilon^{-1} h_*(E \setminus E_2) < \epsilon.$$

This contradicts the requirements for $h_*(E_1)$ and the first part is proved.

Let us now show that $\limsup_{I \to x} h_*(E \cap I)/|h(I)| \geq 1$ for h_*-a.e. point $x \in E$. If not then for some $0 < r < 1$ the set

$$F = \left\{ x \in E : \limsup_{I \to x} \frac{h_*(E \cap I)}{|h(I)|} < r \right\}$$

has positive h_*-measure. Then

$$\beta_3 = \{(I, x) : h_*(E \cap I) \leq r|h(I)|\}$$

is a full covering relation on F; let β_4 be any fine covering relation on F. We may apply the Vitali theorem to the measure h_* (see de Guzmán [3]) to find a packing $\pi \subset \beta_3 \cap \beta_4$ so that

$$h_*(F \setminus \bigcup_{(I,x) \in \pi} I) = 0,$$

and hence

$$h_*(F) \leq \sum_\pi h_*(F \cap I) \leq \sum_\pi h_*(E \cap I) \leq \sum_\pi r|h(I)|.$$

But this gives $h_*(F) \leq r\mathrm{Var}(h, \beta_3 \cap \beta_4)$ and hence that $h_*(F) \leq rh_*(F)$ which is impossible.

Finally we show that $\limsup_{I \to x} h_*(E \cap I)/|h(I)| \leq 1$ for h_*–a.e. point $x \in E$ and the theorem is proved. If not then for some $r > 1$ the set

$$F = \left\{ x \in E : \limsup_{I \to x} \frac{h_*(E \cap I)}{|h(I)|} > r \right\}$$

has positive h_*–measure. Recall by the first part of the proof that the set F_0 of points $x \in F$ for which

$$h_*((E \setminus F) \cap I)/|h(I)| \to 0$$

as $I \to x$ would have $h_*(F \setminus F_0) = 0$. Consequently

$$\limsup_{I \to x} h_*(F \cap I)/|h(I)| > r$$

for all $x \in F_0$. Thus

$$\beta_5 = \{(I, x) : h_*(F \cap I) \geq r|h(I)|\}$$

is a fine covering relation on F_0 and we compute

$$h_*(F) = h_*(F_0) \leq \mathrm{Var}(h, \beta_5) \leq r^{-1}h_*(F)$$

which is impossible and the theorem is proved.

The density theorem for the measures h_* and h^* is intimately related to the identity of the two measures. This is the content of the next theorem. It will allow us to view the identity $h_*(E) = h^*(E)$ as a formalization of a density property.

THEOREM 3.27 *Let h be an interval function and let E be a Borel set of finite h^*–measure. Then the following assertions are equivalent:*

1. $h_*(E) = h^*(E)$.

2. $\lim_{I \to x} h_*(E \cap I)/|h(I)| = 1$ *for* h^**–almost every point* $x \in E$.

3. $\lim_{I \to x} h^*(E \cap I)/|h(I)| = 1$ *for* h^**–almost every point* $x \in E$.

Proof. Let us show that (1) implies (2). Suppose that (1) holds but that (2) fails; then h^* and h_* agree on all h^*–measurable subsets of E and so in view of Theorem 3.26 there is a positive c less than 1 so that

$$\liminf_{I \to x} \frac{h_*(E \cap I)}{|h(I)|} < c$$

on an h^*–measurable set $E' \subset E$ of positive h^*–measure. Let β_1 be the set of pairs (I, x) with $h_*(E \cap I) \le c|h(I)|$. This is a fine covering relation on E' and let β_2 be any full covering relation on this set. Then $\beta_1 \cap \beta_2$ is a fine covering relation on E' so that by the Vitali theorem there is a packing $\pi \subset \beta_1 \cap \beta_2$ so that

$$h_*(E') \le \sum_{(I,x) \in \pi} h_*(E' \cap I) \le \sum_{(I,x) \in \pi} h_*(E \cap I) \le c\mathrm{Var}(h, \beta_2).$$

As β_2 is an arbitrary full covering relation on E' this gives $h_*(E') < h^*(E')$. But this contradicts (1).

Conversely let us show that (2) implies (1). If (2) holds and E_0 is the set of points at which the stated limit does not hold then $h^*(E_0) = 0$. Let $0 < c < 1$ and let β be the collection of pairs (I, x) with $h_*(E \cap I) \ge c|h(I)|$. This is a full covering relation on $E \setminus E_0$ and consequently

$$h^*(E) = h^*(E \setminus E_0) \le \mathrm{Var}(h, \beta) \le c^{-1}h_*(E).$$

Letting $c \to 1$ we deduce $h_*(E) \ge h^*(E)$ and hence $h_*(E) = h^*(E)$ as required.

We now assume that (2) holds. Then by what we have just proved we have the identity $h_*(E \cap I) = h^*(E \cap I)$ for every interval I. Thus (3) must hold trivially.

Finally we show that (3) implies (1). If (3) holds and E_0 is the set of points at which the stated limit does not hold then $h^*(E_0) = 0$. Let β_1 be the collection of pairs (I, x) with $h^*(E \cap I) \le c|h(I)|$. This is a full covering relation on $E \setminus E_0$ for all $c > 1$. Consequently for any fine covering relation β_2 on $E \setminus E_0$ there is a packing $\pi \subset \beta_1 \cap \beta_2$ so that

$$h^*(E) = h^*(E \setminus E_0) \le \sum_{(I,x) \in \pi} h^*(E \cap I) \le c\mathrm{Var}(h, \beta_2).$$

As β_2 is an arbitrary fine covering relation on $E \setminus E_0$ this gives $h_*(E) \geq h^*(E)$ which supplies assertion (1) as required to complete the proof.

4   Derivates

There is a close connection between the derivates of an interval-point function and its variation. We define the derivates in such a way that they relate directly to the notions of full and fine covering relations and this is what provides this connection. Our study relates the derivation properties of interval-point functions to properties of the variational measures. These estimates are used later to characterize the general notions of singularity and absolute continuity for interval-point functions.

4.1   Definitions of the derivates

4.1 DEFINITION. Let h and k be interval-point functions. The derivates of h relative to k are defined as

$$\overline{D}\,(h, k, x) = \lim_{\delta \to 0+} \sup \left\{ |h(I, x)/k(I, x)| : |I| < \delta,\ x \in I^0 \right\}$$

and

$$\underline{D}\,(h, k, x) = \lim_{\delta \to 0+} \inf \left\{ |h(I, x)/k(I, x)| : |I| < \delta,\ x \in I^0 \right\}.$$

Note that these are just the extreme limits of $|h(I_k, x)/k(I_k, x)|$ taken over all sequences of intervals $\{I_k\}$ having x as an interior point and with lengths tending to 0. The quotient $h(I_k, x)/k(I_k, x)$ may fail to exist and rather than exclude this case we merely suppose that $0/0 = 0$ and $c/0 = +\infty$ for $c > 0$.

Most of the concepts developed for interval-point functions relative to the notions of full and fine covering relations are invariant in some sense under the natural equivalence relation $h_1 \equiv h_2$. We shall be able to prove from the material in Section 4.3 below that if $h_1 \equiv h_2$ then, while $\overline{D}\,(h_1, k, x)$ and $\overline{D}\,(h_2, k, x)$ may differ, it is nonetheless true that

$$\overline{D}\,(h_1, k, x) = \overline{D}\,(h_2, k, x)$$

and

$$\underline{D}\,(h_1, k, x) = \underline{D}\,(h_2, k, x)$$

at k_*–almost every point x.

4.2   Baire class of derivates

It is essential that we establish the measurability of these extreme derivates. More generally we can as easily determine the Baire classification. Recall that an ordinary derivative f' can be seen to be measurable from the expression

$$f'(x) = \lim_{n \to \infty} n(f(x + 1/n) - f(x)),$$

exhibiting f' as a pointwise limit of a sequence of continuous functions. As well as providing the measurability this places f' in the first Baire class which is important to an understanding of the structure of derivatives. There is now a considerable literature devoted to establishing the precise Baire classification of a variety of generalized derivatives.

For our purposes we mainly wish to know that the extreme derivates are Baire functions and hence are measurable. They belong to the second Baire class: they are pointwise limits of functions in the first Baire class. This is equivalent to a determination of the Borel structure of the associated sets as we see in Theorem 4.2. This can be considered as an analogue of a theorem of Hájek (see, for example, [2, Theorem 2.3, p. 57]). Similar considerations are given in [15, pp. 152-158] for derivates of a nondecreasing function f that are of the form

$$\limsup_{y,z \to x,\ y<x<z} \frac{f(z) - f(y)}{\phi(y - z)}.$$

These same methods apply to yield our theorem.

THEOREM 4.2 *Let h and k be interval functions. Then the function $\overline{D}(h, k, x)$ is in the second Baire class. In particular for any positive number t the set*

$$\left\{ x : \overline{D}(h, k, x) \geq t \right\}$$

is a Borel set of type $\mathcal{G}_\delta$ and the set

$$\left\{ x : \overline{D}(h, k, x) > t \right\}$$

is a Borel set of type $\mathcal{G}_{\delta\sigma}$.

Proof. The set $\left\{ x : \overline{D}(h, k, x) \geq t \right\}$ may be expressed as

$$\bigcap_{n=1}^{\infty} \bigcup \left\{ I^0 : I \in S_n \right\}$$

where S_n denotes the collection of all intervals I for which $|I| < 1/n$ and $|h(I)/k(I)| \geq t - 1/n$. This is clearly a $\mathcal{G}_\delta$ set.

The identity

$$\left\{x : \overline{D}\,(h, k, x) > t\right\} = \bigcup_{n=1}^{\infty} \left\{x : \overline{D}\,(h, k, x) \geq t + 1/n\right\}$$

then completes the proof.

Of course the same considerations apply to the lower derivate $\underline{D}\,(h, k, x)$.

4.3 Variational estimates

We begin with elementary estimates that relate the upper and lower derivates of an interval-point function h relative to an interval-point function k to the variational measures h^*, k^*, h_* and k_*. In each case there are dual estimates available for the upper and lower derivates in terms of a different pair of measures.

THEOREM 4.3 *If* $\overline{D}\,(h, k, x) \geq C$ *at every point* x *of a set* E *then*

$$C k_*(E) \leq h^*(E).$$

If $\underline{D}\,(h, k, x) \leq C$ *at every point* x *of a set* E *then*

$$C k^*(E) \geq h_*(E).$$

Proof. Under this assumption on E the collection

$$\beta_1 = \{(I, x) : |h(I, x)| \geq C'|k(I, x)|\}$$

is a fine covering relation on E for any $C' < C$. For any full covering relation β_2 on E the intersection $\beta_1 \cap \beta_2$ is a fine covering relation on E and hence

$$C' k_*(E) \leq C'\mathrm{Var}(k, \beta_1 \cap \beta_2) \leq \mathrm{Var}(h, \beta_2).$$

Since β_2 is arbitrary and $C' < C$ is arbitrary this gives $C k_*(E) \leq h^*(E)$ as required. The second part of the assertion is the dual and has a similar proof obtained by interchanging the roles of full and fine covers.

THEOREM 4.4 *If* $\overline{D}(h, k, x) \leq C$ *at every point* x *of a set* E *then*

$$h_*(E) \leq C k_*(E) \ and \ h^*(E) \leq C k^*(E).$$

If $\underline{D}(h, k, x) \geq C$ *at every point* x *of a set* E *then*

$$h^*(E) \geq C k^*(E) \ and \ h_*(E) \geq C k_*(E).$$

Proof. Under this assumption on E the collection

$$\beta_1 = \{(I, x) : |h(I, x)| \leq C'|k(I, x)|\}$$

is a full covering relation on E for any $C' > C$. For any full covering relation β_2 on E the intersection $\beta_1 \cap \beta_2$ is a full covering relation on E (by Lemma 2.6) and hence

$$h^*(E) \leq \mathrm{Var}(h, \beta_1 \cap \beta_2) \leq C'\mathrm{Var}(k, \beta_2).$$

Since β_2 is arbitrary and $C' > C$ is arbitrary this gives $h^*(E) \leq C k^*(E)$. Again for any fine covering relation β_3 on E the intersection $\beta_1 \cap \beta_3$ is a fine covering relation on E (by Lemma 2.7) and hence

$$h_*(E) \leq \mathrm{Var}(h, \beta_1 \cap \beta_3) \leq C'\mathrm{Var}(k, \beta_3).$$

As before this gives $h_*(E) \leq C k_*(E)$ as required. The remaining assertion permits a dual argument.

From these elementary estimates we draw several corollaries.

COROLLARY 4.5 *If* $\overline{D}(h, k, x) = \infty$ *at every point* x *of a set* E *then* $k_*(E \cap M) = 0$ *for every set* M *on which* h^* *is* σ–*finite. If* $\underline{D}(h, k, x) = \infty$ *at every point* x *of a set* E *then* $k^*(E \cap M) = 0$ *for every set* M *on which* h^* *is* σ–*finite and* $k_*(E \cap M) = 0$ *for every set* M *on which* h_* *is* σ–*finite.*

Proof. This follows from the inequality in Theorem 4.3.

COROLLARY 4.6 *If* $\overline{D}(h, k, x) = 0$ *at every point* x *of a set* E *then* $h^*(E \cap M) = 0$ *for every set* M *on which* k^* *is* σ–*finite and* $h_*(E \cap N) = 0$ *for every set* N *on which* k_* *is* σ–*finite. If* $\underline{D}(h, k, x) = 0$ *at every point* x *of a set* E *then* $h_*(E \cap M) = 0$ *for every set* M *on which* k^* *is* σ–*finite.*

Proof. This is an immediate consequence of the inequalities in Theorem 4.4.

COROLLARY 4.7 *Let h and k be interval-point functions and let E be a set of points such that $0 < \overline{D}(h,k,x) < \infty$ for each $x \in E$. Then*

$$h_*(E) \leq (fk)_*(E) \leq h^*(E)$$

where f denotes the function $x \to \overline{D}(h,k,x)$. Let F be a set of points such that $0 < \underline{D}(h,k,x) < \infty$ for each $x \in F$. Then

$$h_*(F) \leq (gk)^*(F) \leq h^*(F)$$

where g denotes the function $x \to \underline{D}(h,k,x)$.

Proof. We note that the relation $\overline{D}(h,k,x) = f(x)$ is exactly equivalent to the assertion that $\overline{D}(h,fk,x) = 1$. These inequalities then are immediate consequences of the inequalities in Theorems 4.3 and 4.4.

COROLLARY 4.8 *Let h and k be interval-point functions and let E be a set of points. Then*

$$\frac{h_*(E)}{\sup_{x \in E}\overline{D}(h,k,x)} \leq k_*(E) \leq \frac{h^*(E)}{\inf_{x \in E}\overline{D}(h,k,x)}$$

and

$$\frac{h_*(E)}{\sup_{x \in E}\underline{D}(h,k,x)} \leq k^*(E) \leq \frac{h^*(E)}{\inf_{x \in E}\underline{D}(h,k,x)}.$$

Proof. This is just a variant of Theorem 4.7.

4.4   Lipschitz conditions

Classically estimates on the derivates of a function are related to Lipschitz conditions. Thus, for a well-known example, a function with bounded Dini derivatives on an interval must be Lipschitz on that interval. Our results here in this setting are a bit more subtle and require different techniques.

4.9 DEFINITION. Let h and k be interval-point functions and suppose that E is a set of real numbers. We say that h is *uniformly Lip(k)* on the set E provided there is a positive number δ such that the inequality

$$|h(I,x)| \leq C|k(I,x)|$$

holds for every interval-point pair (I,x) for which x belongs to E, x is an interior point of I and $|I| < \delta$.

We require on occasion a stronger kind of Lipschitz condition; note the use of the lower case "l" as in the Landau big-O, little-o notation (we are following [15, p. 149] here).

4.10 DEFINITION. Let h and k be interval-point functions and suppose that E is a set of real numbers. We say that h is *uniformly lip(k)* on the set E provided that for every positive number ϵ there is a positive number δ such that the inequality

$$|h(I,x)| \leq \epsilon|k(I,x)|$$

holds for every interval-point pair (I,x) for which x belongs to E, x is an interior point of I and $|I| < \delta$.

Our results show that the sets of points on which the derivates are finite (zero) can be arranged as a union of sets on which there is a uniform Lip(k) (lip(k)) condition.

THEOREM 4.11 *Let h and k be interval-point functions and let E be the set of points x at which $\overline{D}(h,k,x) < \infty$. Then E permits a partition $E = \bigcup_{n=1}^{\infty} E_n$ in such a way that h is uniformly Lip(k) on each set E_n.*

Proof. Let $A_n = \left\{ x \in E : \overline{D}(h,k,x) < n \right\}$ and choose a full covering relation β on A_n by setting

$$\beta = \{(I,x) : |h(I,x)| \leq n|k(I,x)|\}.$$

Then by the covering Lemma 2.11 there is an increasing sequence $\{A_{nk}\}$ of sets whose union is A_n and each pair (I,x) with $|I| < 1/k$ and $x \in A_{nk} \cap I^0$ belonging to β. But this says precisely that h is Lip(k) on A_{nk}. Now

$$E = \bigcup_{n=1}^{\infty} \bigcup_{k=1}^{\infty} A_{nk}$$

so that it remains only to arrange these sets into a single disjoint sequence and the theorem is proved.

If h and k are interval functions then the result in Theorem 4.11 can be made somewhat more precise.

THEOREM 4.12 *Let h and k be interval functions and let E be the set of all points x such that $\overline{D}(h, k, x) < \infty$. Then E permits a Borel partition $E = \bigcup_{n=1}^{\infty} E_n$ in such a way that h is uniformly $Lip(k)$ on the closure of each set E_n $(n \geq 1)$.*

Proof. By Theorem 4.2 the set E is a Borel set and then Theorem 4.11 supplies a partition $\{A_k\}$. Because h and k are interval functions this extends the $\text{Lip}(k)$ property to the closures $\{\overline{A}_k\}$. This then allows the partition to be adjusted so as to be formed of Borel sets: take $E_1 = E \cap \overline{A}_1$, $E_2 = E \cap \overline{A}_2 \setminus E_1$ and so on.

The next theorem is a sharper version of Theorem 4.12 for zero derivatives.

THEOREM 4.13 *Let h and k be interval functions so that h^* is σ-finite and let E be the set of points x at which $\overline{D}(h, k, x) = 0$. Then E permits a Borel partition $E = \bigcup_{n=0}^{\infty} E_n$ in such a way that $h^*(E_0) = 0$ and h is uniformly $lip(k)$ on each set E_n.*

Proof. (cf. [15, pp. 164–166].) We may suppose that h^* is finite. The set of points E at which $\overline{D}(h, k, x) = 0$ may be displayed as

$$E = \bigcap_{m=1}^{\infty} \bigcup_{n=1}^{\infty} A_{mn}$$

where A_{mn} is the set of all points x with the property that $|h(I)/k(I)| < 1/m$ for any interval I which contains x as an interior point and which has length less than $1/n$. These are evidently Borel sets of finite h^*-measure. By a measure–theoretic version of Egorov's theorem (see [15, p. 39]) there is an increasing sequence of indices $\{n_k\}$ so that for each p=1,2,...

$$h^*(B_p) \geq h^*(E) - 2^{-p+1} \tag{15}$$

where

$$B_p = \bigcap_{m=p}^{\infty} \bigcup_{n=1}^{n_k} A_{mn} = \bigcap_{m=p}^{\infty} A_{mn_k}.$$

The sequence $\{B_p\}$ is composed of Borel sets which, because of the inequality (15) above, covers all of E except for a set of h^*–measure zero. Set $E_i = B_i \setminus \bigcup_{j<i} B_j$ for $i \geq 1$ and set $E_0 = E \setminus \bigcup_{j \geq 1} B_j$. This sequence satisfies each of the required conditions in the assertion of the theorem and the proof is complete. For example if $\epsilon > 0$ and i is an arbitrary index then choose $N > i$ so that $1/N < \epsilon$; then any interval I with length less than $1/n(N)$ and whose interior contains a point of E_i must contain a point of $A_{Nn(N)}$ which requires that $|h(I)/k(I)| < 1/N$ and hence that $|h(I)| \leq \epsilon|k(I)|$ as required.

We should remark that in Theorem 4.13 should k^* be σ–finite then the entire set E has h^*–measure zero and so the partition is trivial. Thus the only applicability of this result is in the non σ–finite case.

4.5   Exact derivatives

Exact derivatives as well as extreme derivates may be introduced in this setting. Let h and k be arbitrary interval-point functions. We say that the *derivative* of h with respect to k at a point x is the real number c provided $\overline{D}(h - ck, k, x) = 0$. In symbols we write $D(h, k, x) = c$. Note that the relation $D(h, k, x) = f(x)$ holds at every point of a set E if and only if $\overline{D}(h - fk, k, x) = 0$ at every point in E. Note also that the exact derivative need not be unique since h and k may vanish; generally in applications this is no problem.

The following observations give relations between the variation and the exact derivative. The proofs follow easily from the propositions in Section 4.3 and are just restatements of those propositions in the language of exact derivatives. Note first that a function with zero variation on a set has a zero derivative outside of a "small" set: if $h^*(E) = 0$ then $D(h, k, x) = 0$ at k_*–almost every point in E. Similarly a function with finite variation on a set has finite derivatives except on a small set: if $k^*(E) < \infty$ and $D(h, k, x) = 0$ at k^*–almost every point in E then $h^*(E) = 0$. Finally there is a version of the fundamental theorem of the calculus in this setting: if $k^*(E) < \infty$ and $D(h, k, x) = f(x)$ at h^*–almost every and at k^*–almost every point in E then

$h \equiv fk$ on E. If $h \equiv fk$ on E then $D(h, k, x) = f(x)$ at k_*–almost every point in E.

5   Absolute continuity and singularity

The classical notions of absolute continuity and singularity for point functions will be generalized in this section so as to apply to any interval-point functions. We shall use the convenient symbolic notations

$$h_1 \ll h_2 \quad \text{and} \quad h_1 \perp h_2$$

to indicate these notions and mostly avoid the expressions "absolute continuity" and "singularity" except as they agree with their classical usage. Together with the equivalence relation $h_1 \equiv h_2$ these three relations between pairs of interval-point functions are central to the theory. Loosely the equivalence relation $h_1 \equiv h_2$ means that h_1 and h_2 have nearly identical properties in so far as this differentiation basis is concerned; the partial order $h_1 \ll h_2$ means that h_1 is "smaller" than h_2 so that h_1 inherits a number of properties of h_2; and finally the relation $h_1 \perp h_2$ represents a kind of mutual orthogonality. The notational intention is the same as that used in many presentations of measure theory and we shall see that the expected measure properties for the associated measures will hold. A minor unpleasantness that comes from accepting symbolic expressions of ideas should in this case, we hope, be overcome by the ease with which certain ideas can be expressed and manipulated.

For a real function f defined on an interval $[a, b]$ the notion of absolute continuity is defined by requiring that the sum $\sum_{i=1}^{n} |f(b_i) - f(a_i)|$ can be made arbitrarily small provided $\{[a_i, b_i]\}$ is a sequence of nonoverlapping subintervals of $[a, b]$ of sufficiently small total length. The parallel notion of singularity requires that for sequences $\{[a_i, b_i]\}$ of nonoverlapping subintervals of $[a, b]$ there is a split of the indices $\{1, 2, 3, \ldots, n\} = A \cup B$ in such a way that the separate sums $\sum_{i \in A} |f(b_i) - f(a_i)|$ and $\sum_{i \in B} |b_i - a_i|$ are small.

These same ideas may be carried over to the study of the integration and differentiation properties of interval-point functions. Note that our definitions are direct generalizations of the classical ones but have a broader range of application. Note too that the definitions are tied to the notions of full and fine covering relations and so are dependent on the differentiation basis underlying the theory.

5.1   Basic Definitions

There are two dual versions of absolute continuity and two of mutual singularity.

5.1 DEFINITION. Let h and k be interval-point functions and let E be a set of real numbers. We write $h \ll k$ on E provided that for every $\epsilon > 0$ there is a $\delta > 0$ and a full covering relation β on E so that whenever $\pi \subset \beta$ is a packing for which

$$\sum_{(I,x)\in\pi} |k(I,x)| < \delta$$

then necessarily

$$\sum_{(I,x)\in\pi} |h(I,x)| < \epsilon.$$

A weaker form of absolute continuity uses fine covers.

5.2 DEFINITION. Let h and k be interval-point functions and let E be a set of real numbers. We write $h \ll k$ weakly on E provided that for every $\epsilon > 0$ there is a $\delta > 0$ and a fine covering relation β on E so that whenever $\pi \subset \beta$ is a packing for which

$$\sum_{(I,x)\in\pi} |k(I,x)| < \delta$$

then necessarily

$$\sum_{(I,x)\in\pi} |h(I,x)| < \epsilon.$$

The mutual singularity is defined in a similar way, also in dual versions.

5.3 DEFINITION. Let h and k be interval-point functions and let E be a set of real numbers. We write $h \perp k$ on E provided that for every $\epsilon > 0$ there is a full covering relation β on E so that whenever $\pi \subset \beta$ is a packing there is a decomposition $\pi = \pi_1 \cup \pi_2$, with $\pi_1 \cap \pi_2 = \emptyset$, for which

$$\sum_{(I,x)\in\pi_1} |k(I,x)| < \epsilon$$

and
$$\sum_{(I,x)\in\pi_2} |h(I,x)| < \epsilon.$$

5.4 DEFINITION. Let h and k be interval-point functions and let E be a set of real numbers. We write $h \perp k$ weakly on E provided that for every $\epsilon > 0$ there is a fine covering relation β on E so that whenever $\pi \subset \beta$ is a packing there is a decomposition $\pi = \pi_1 \cup \pi_2$, with $\pi_1 \cap \pi_2 = \emptyset$, for which

$$\sum_{(I,x)\in\pi_1} |k(I,x)| < \epsilon$$

and

$$\sum_{(I,x)\in\pi_2} |h(I,x)| < \epsilon.$$

Our first observation is that the concepts are preserved by the equivalence relation $h \equiv k$ in the sense that if $h_1 \equiv h_2$ on a set E and $k_1 \equiv k_2$ on E then the expressions $h_1 \ll k_1$ on E and $h_2 \ll k_2$ on E are equivalent. So too are the expressions $h_1 \perp k_1$ on E and $h_2 \perp k_2$ on E.

THEOREM 5.5 *Let h_1, h_2, k_1 and k_2 be interval-point functions and E a set of real numbers such that $h_1 \equiv h_2$ on E and $k_1 \equiv k_2$ on E. Then $h_1 \perp k_1$ on E if and only if $h_2 \perp k_2$ on E and $h_1 \ll k_1$ on E if and only if $h_2 \ll k_2$ on E.*

Proof. Let $\epsilon > 0$. Choose $\delta > 0$ and a full covering relation β_1 on E so that for any packing $\pi \subset \beta_1$ if $\sum_{(I,x)\in\pi} |k_1(I,x)| < \delta$ then necessarily

$$\sum_{(I,x)\in\pi} |h_1(I,x)| < \epsilon/2. \tag{16}$$

Choose full covering relations β_2 and β_3 on E so that

$$\text{Var}(h_1 - h_2, \beta) < \epsilon/2 \tag{17}$$

and

$$\text{Var}(k_1 - k_2, \beta) < \delta/2. \tag{18}$$

Then we define the full covering relation β on E as $\beta = \beta_1 \cap \beta_2 \cap \beta_3$. For any packing $\pi \subset \beta$ for which $\sum_{(I,x)\in\pi} |k_2(I,x)| < \delta/2$ we will have, because of the inequality (18), $\sum_{(I,x)\in\pi}|k_1(I,x)| < \delta$ and hence by the inequalities (16) and (17) that $\sum_{(I,x)\in\pi}|h_2(I,x)| < \epsilon$. This verifies the condition of the definition and proves the first part of the theorem. The second part has a similar proof.

We list now some elementary properties of these notions. As most of these may be proved either easily or trivially from the definitions together with our standard properties of the variation we omit the proofs. Each relation is intended to hold on some set E and accordingly reference to that set may be suppressed. First note that the relation "$\ll$" is transitive as the symbols would suggest: if $h_1 \ll h_2$ and $h_2 \ll h_3$ then $h_1 \ll h_3$. There are some simple arithmetic properties: if $h_1 \ll h_3$ and $h_2 \ll h_3$ then $c_1 h_1 + c_2 h_2 \ll h_3$; if $h_1 \perp h_3$ and $h_2 \perp h_3$ then $c_1 h_1 + c_2 h_2 \perp h_3$; if $h_1 + h_2 \ll h_3$ where $h_1 \geq 0$ and $h_2 \geq 0$ then both $h_1 \ll h_3$ and $h_2 \ll h_3$.

For any interval-point function h the relations $h \ll h$, $0 \ll h$ and $h \perp 0$ are true. The relation $h \perp h$ holds if and only if $h \equiv 0$. Finally for any interval-point functions h and k both the relations $h \ll k$ and $h \perp k$ can hold if and only if $h \equiv 0$.

5.2   Further properties

In addition to the immediate and trivial properties of these notions we shall need some deeper ones. If the relations $h \ll k$ or $h \perp k$ hold on each E_i from a sequence of sets $\{E_i\}$ then we will need to know if the relation may be extended to the union of the sets in the sequence. Here we give two positive results in this direction.

The analogues for the weak notions may be similarly obtained.

THEOREM 5.6 *Let h and k be a pair of interval-point functions such that the relation $h \perp k$ on E_i holds for each member of a sequence of sets $\{E_i\}$ and let $E = \bigcup_{i=1}^{\infty} E_i$. Then $h \perp k$ on E.*

Proof.   We may suppose that the sets $\{E_i\}$ are disjoint. Let $\epsilon > 0$ and, for each integer i, choose $\delta_i > 0$ and a full covering relation β_i on E_i so that any packing $\pi_i \subset \beta_i$ may be decomposed into a pair $\{\pi_{i1}, \pi_{i2}\}$ for which the inequalities

$$\sum_{(I,x)\in\pi_{i1}} |h(I,x)| < \epsilon/2^{i+1} \tag{19}$$

and

$$\sum_{(I,x)\in\pi_{i2}} |k(I,x)| < \epsilon/2^{i+1} \tag{20}$$

hold.

We write $\beta = \bigcup_{i=1}^{\infty} \beta_i[E_i]$ and observe that β is a full covering relation on E. For any packing π from β we write $\pi_i = \pi[E_i]$ and then use the above decompositions to obtain π_{i1} and π_{i2} so that the inequalities (19) and (20) hold. Set $\pi_1 = \bigcup \pi_{i1}$ and $\pi_2 = \bigcup \pi_{i2}$. This then gives the inequalities

$$\sum_{(I,x)\in\pi_1} |h(I,x)| < \epsilon$$

and

$$\sum_{(I,x)\in\pi_2} |k(I,x)| < \epsilon$$

which we require in order to establish the desired relation $h \perp k$ on E.

THEOREM 5.7 *Let h and k be a pair of interval-point functions such that the relation $h \ll k$ on E_i holds for each member of a sequence of sets $\{E_i\}$ and let $E = \bigcup_{i=1}^{\infty} E_i$. Suppose that each set E_i is h^*-measurable and that $h^*(E) < \infty$. Then $h \ll k$ on E.*

Proof. We may suppose that the sets $\{E_i\}$ are disjoint. Let $\epsilon > 0$ and choose, for each integer i, $\delta_i > 0$ and a full covering relation β_i on E_i so that for any packing $\pi \subset \beta_i$ if $\sum_{(I,x)\in\pi} |k(I,x)| < \delta_i$ then necessarily

$$\sum_{(I,x)\in\pi} |h(I,x)| < \epsilon/2^{i+1}. \tag{21}$$

Because the sets involved are h^*-measurable and h^* is finite on E we may choose an integer N so that

$$h^*\left(\bigcup_{i=N+1}^{\infty} E_i\right) < \epsilon/2. \tag{22}$$

Because of the inequality (22) there is a full covering relation β_0 on the set $E_0 = \bigcup_{i=N+1}^{\infty} E_i$ so that

$$\mathrm{Var}(h, \beta_0) < \epsilon/2. \tag{23}$$

We write $\beta = \bigcup_{i=1}^{N} \beta_i[E_i] \cup \beta_0[E_0]$ and observe that β is a full covering relation on E. Set $\delta = \min\{\delta_1, \delta_2, \ldots, \delta_N\}$. Now if π is any packing contained in β for which $\sum_{\pi} |k(I, x)| < \delta$ then $\sum_{\pi[E \setminus E_0]} |h(I, x)| < \epsilon/2$ because of the inequality (21) and $\sum_{\pi[E]} |h(I, x)| < \epsilon/2$ because of the inequality (23). Thus $\sum_{\pi} |h(I, x)| < \epsilon$ as required to establish the relation $h \ll k$ on the set E.

We conclude with one more extension theorem. If $h \ll k$ then the product fh for a point function f will satisfy $fh \ll k$ under certain hypotheses.

THEOREM 5.8 *Let h and k be a pair of interval-point functions such that the relation $h \ll k$ on E holds. Suppose that the point function f is $(fh)^*$–measurable, that E is $(fh)^*$–measurable and that $(fh)^*(E) < \infty$. Then $fh \ll k$ on E.*

Proof.   Let $\epsilon > 0$. Define the sets $E_n = \{x \in E : |f(x)| > n\}$. Then the sets $\{E_n\}$ are decreasing to $\emptyset$ and are $(fh)^*$–measurable so that $(fh)^*(E_n) \to 0$ as $n \to \infty$. Choose N so large that $(fh)^*(E_N) < \epsilon/2$ and obtain a full covering relation β_N on E_N in such a way that $\text{Var}(fh, \beta_N) < \epsilon/2$.

Since $h \ll k$ we may choose a full covering relation γ on E and a positive number δ so that whenever $\pi \subset \gamma$ with $\sum_{\pi} |k(I, x)| < \delta$ then necessarily $\sum_{\pi} |h(I, x)| < \epsilon/2N$.

Write $\beta = \gamma[E \setminus E_N] \cup \beta_N$. This is a full covering relation on E and the following computations show that $fh \ll k$: if π is a packing from β for which $\sum_{\pi} |k(I, x)| < \delta$ then write $\pi_1 = \pi[E_N]$ and $\pi_2 = \pi[\mathbb{R} \setminus E_N]$ and compute

$$
\begin{aligned}
\sum_{\pi} |f(x)h(I, x)| &= \sum_{\pi_1} |f(x)h(I, x)| + \sum_{\pi_2} |f(x)h(I, x)| \\
&\leq \text{Var}(fh, \beta_N) + \sum_{\pi_2} N|h(I, x)| \\
&< \epsilon/2 + N(\epsilon/2N) = \epsilon.
\end{aligned}
$$

5.3   Measure properties

Evidently relations such as $h \ll k$ and $h \perp k$ must impose some relationships between the pair of variational measures h^* and k^* and the pair h_* and k_*. Our first assertion shows that this relation imposes absolute continuity requirements on these measures.

THEOREM 5.9 *Suppose that the relation $h \ll k$ on E holds for interval-point functions h and k. Then for every $\epsilon > 0$ there is a $\delta > 0$ so that whenever $A \subset E$ and $k^*(A) < \delta$ then necessarily $h^*(A) < \epsilon$, and so that whenever $A \subset E$ and $k_*(A) < \delta$ then necessarily $h_*(A) < \epsilon$.*

Proof. Let $\epsilon > 0$. Choose $\delta > 0$ and a full covering relation β_1 on E so that for any packing $\pi \subset \beta_1$ if $\sum_{(I,x) \in \pi} |k(I,x)| < \delta$ then necessarily

$$\sum_{(I,x) \in \pi} |h(I,x)| < \epsilon. \tag{24}$$

Now if A is a subset of E with $k^*(A) < \delta$ there is a full covering relation β_2 on E so that

$$\mathrm{Var}(k, \beta_2) < \delta \tag{25}$$

also. By the inequality (24) above $h^*(A) \leq \mathrm{Var}(h, \beta_1 \cap \beta_2) < \epsilon$ as required.

The proof for the fine variations is almost identical except that in this case β_2 would be chosen as a fine covering relation on A for which the inequality (25) holds.

A dual version of Theorem 5.9 can be similarly proved.

THEOREM 5.10 *Suppose that the relation $h \ll k$ weakly on E holds for interval-point functions h and k. Then for every $\epsilon > 0$ there is a $\delta > 0$ so that whenever $A \subset E$ and $k^*(A) < \delta$ then necessarily $h_*(A) < \epsilon$.*

Familiar measure-theoretic consequences are now easy to obtain. For example suppose that the relation $h \ll k$ on E holds for interval-point functions h and k. Then $h^*(A \cap E)$ vanishes whenever $k^*(A)$ vanishes and $h_*(A \cap E)$ vanishes whenever $k_*(A)$ vanishes. This follows directly from Theorem 5.9. Similarly if the relation $h \ll k$ weakly on E holds for interval-point functions h and k then $h_*(A \cap E)$ vanishes whenever $k^*(A)$ vanishes. This follows directly from Theorem 5.10. Note the continuity requirements of these assertions: if the relation $h \ll k$ on E holds for interval-point functions h and k then h is continuous at any point $x \in E$ at which k is continuous and is weakly continuous at any such point at which k is weakly continuous. This follows directly from the preceding comments since the continuity of a interval-point function h at a point x depends only on the vanishing of the measure $h^*(\{x\})$, and the weak continuity of h at x depends only on the vanishing of the measure $h_*(\{x\})$.

Orthogonality relations too have measure-theoretic versions. Two measures are considered orthogonal on a set E if they are concentrated on disjoint subsets of E; more precisely if μ and ν are outer measures then an orthogonality relation on a set E might mean that E can be decomposed into two disjoint subsets E_1 and E_2 so that $\mu(E_1) = \nu(E_2) = 0$. Again we see in Theorem 5.11 that this condition is stronger than the relation $h \perp k$ and later we give conditions under which they are equivalent.

THEOREM 5.11 *Let h and k be interval-point functions and suppose that a set E of real numbers may be decomposed into two disjoint subsets E_1 and E_2 so that $h^*(E_1) = k^*(E_2) = 0$. Then $h \perp k$ on E.*

Proof. Let $\epsilon > 0$. Then there are full covering relations β_1 and β_2 on E_1 and E_2 respectively so that $\mathrm{Var}(h, \beta_1) < \epsilon$ and $\mathrm{Var}(k, \beta_2) < \epsilon$. The collection $\beta = \beta_1[E_1] \cup \beta_2[E_2]$ is a full covering relation on E. If π is any packing contained in β the decomposition $\pi_1 = \pi[E_1]$ and $\pi_2 = \pi[E_2]$ evidently satisfies the inequalities

$$\sum_{\pi_1} |h(I, x)| \le \mathrm{Var}(h, \beta_1) < \epsilon$$

and

$$\sum_{\pi_2} |h(I, x)| \le \mathrm{Var}(h, \beta_2) < \epsilon$$

as required to establish the relation $h \perp k$ on E.

5.4   A stronger orthogonality relation

The relation $h \perp k$ may, under certain additional hypotheses, be written in other equivalent forms. In this section we introduce a further notion that has connections with this "singularity" idea. We investigate the relation $\sqrt{|hk|} \equiv 0$ on E; by this we mean, of course, that

$$V^*\left(\sqrt{|hk|}, E\right) = 0.$$

Our first observation in Theorem 5.12 is that this condition is stronger than the relation $h \perp k$; later in Theorem 5.13 we give conditions under which they are equivalent.

THEOREM 5.12 *For any interval-point functions h and k the relation $\sqrt{|hk|} \equiv 0$ on a set E requires that $h \perp k$ on E.*

Proof. Let $\epsilon > 0$ and suppose that $V^* \left(\sqrt{|hk|}, E \right) = 0$. Then there is a full covering relation β on E so that $\mathrm{Var} \left(\sqrt{|hk|}, \beta \right) < \epsilon$. If π is any packing contained in β then we can exhibit the decomposition of π required by Definition 5.3 as follows: we write

$$\pi_1 = \{ (I, x) \in \pi : |h(I, x)| \leq |k(I, x)| \} \tag{26}$$

and

$$\pi_2 = \{ (I, x) \in \pi : |h(I, x)| > |k(I, x)| \}. \tag{27}$$

Certainly $\pi = \pi_1 \cup \pi_2$ and $\pi_1 \cap \pi_2 = \emptyset$. Further the inequality in (26) shows that

$$\sum_{\pi_1} |h(I, x)| \leq \sqrt{|h(I, x) k(I, x)|} \leq \mathrm{Var} \left(\sqrt{|hk|}, \beta \right) < \epsilon.$$

In the same way we obtain $\sum_{\pi_2} |k(I, x)| < \epsilon$ from (27) and this establishes that the decomposition of π has the required properties and so we have proved that $h \perp k$ on E.

THEOREM 5.13 *Let h and k be interval-point functions, let E be a set of real numbers and suppose that h^* and k^* are σ–finite on E. Then the relation $\sqrt{|hk|} \equiv 0$ on E holds if and only if $h \perp k$ on E.*

Proof. Because of Theorem 5.12 we need prove only one direction. We suppose that E has finite h^* and k^* measures and we obtain a proof by showing that for any $\epsilon > 0$ the relation $h \perp k$ on a set E allows us to choose a full covering relation β on E in such a way that

$$\mathrm{Var} \left(\sqrt{|hk|}, \beta \right) < M \sqrt{\epsilon} \tag{28}$$

for a suitable constant M.

There are full covering relations β_1 and β_2 such that

$$\mathrm{Var}(h, \beta_1) < h^*(E) + 1$$

and

$$\mathrm{Var}(k, \beta_2) < k^*(E) + 1.$$

Because $h \perp k$ on the set E we may choose a full covering relation β_3 on E so that the decomposition of any $\pi \subset \beta_3$ as in Definition 5.3 is allowed; thus we will have the inequalities

$$\sum_{(I,x)\in\pi_1} |h(I,x)| < \epsilon \tag{29}$$

and

$$\sum_{(I,x)\in\pi_2} |k(I,x)| < \epsilon \tag{30}$$

are available.

Define $\beta = \beta_1 \cap \beta_2 \cap \beta_3$. This is a full covering relation too on E and we will show that the inequality (28) holds for this β. If π is any packing contained in β there is a decomposition π_1 and π_2 so that the inequalities (29) and (30) hold. Using these and the standard Cauchy-Schwartz inequality we have

$$\sum_\pi \sqrt{|h(I,x)k(I,x)|}$$
$$= \sum_{\pi_1} \sqrt{|h(I,x)k(I,x)|} + \sum_{\pi_2} \sqrt{|h(I,x)k(I,x)|}$$
$$\leq \sqrt{\sum_{\pi_1}|h(I,x)|}\sqrt{\sum_{\pi_1}|k(I,x)|} + \sqrt{\sum_{\pi_1}|h(I,x)|}\sqrt{\sum_{\pi_1}|k(I,x)|}$$
$$\leq \sqrt{\epsilon(k^*(E)+1)} + \sqrt{\epsilon(h^*(E)+1)}.$$

As this inequality holds for all packings $\pi \subset \beta$ we have the inequality (28).

The theorem is proved then on sets E with finite h^* and k^* measure. By the measure properties of the variation this then extends to E since E may be written as a countable union of sets $\{E_i\}$ on each of which the equation $V^*\left(\sqrt{|hk|}, E_i\right) = 0$ holds.

5.5   Derivation properties and singularity

The classical notion of a singular function of bounded variation may be defined by a variational procedure (as discussed in Section 5.1) or in terms of its derivative properties. Thus a function f of bounded variation on an interval $[a, b]$ is singular if and only if $f'(x) = 0$ almost everywhere there. We give some general versions of this.

THEOREM 5.14 *Let h and k be interval-point functions. Suppose that k^* is σ–finite on a set E of real numbers and that at k^*–almost every point $x \in E$ the derivative $\overline{D}(h, k, x) = 0$. Then the set E may be decomposed into two disjoint subsets E_1 and E_2 so that $h^*(E_1) = k^*(E_2) = 0$ and consequently $h \perp k$ on E.*

Proof. If E_1 denotes the set of points $x \in E$ at which the stated derivative exists and vanishes then we know by Theorem 4.6 that $h^*(E_1) = 0$. But by hypothesis this derivative vanishes k^*–almost everywhere in E and so $k^*(E \setminus E_1) = 0$ which is the required decomposition of E. The final assertion of the theorem follows by an application of 5.11.

The next theorem is similar.

THEOREM 5.15 *Let h and k be interval-point functions. Suppose that h^* is σ–finite on a set E of real numbers and that at h^*–almost every point $x \in E$ the derivative $\underline{D}(h, k, x) = +\infty$. Then the set E may be decomposed into two disjoint subsets E_1 and E_2 so that $h^*(E_1) = k^*(E_2) = 0$ and consequently $h \perp k$ on E.*

Proof. Let E_1 denote the set of points $x \in E$ at which $\underline{D}(h, k, x) = +\infty$ and let $E_2 = E \setminus E_1$. Then $h^*(E_2) = 0$ and, since $\underline{D}(h, k, x) = +\infty$ at every point $x \in E_1$, Theorem 4.5 shows that $k^*(E_1) = 0$.

In the converse direction we have the following results.

THEOREM 5.16 *If $h \perp k$ on a set E then $\underline{D}(h, k, x) = 0$ at k^*–almost every point in E, $\underline{D}(h, k, x) = +\infty$ at h_*–almost every point in E, $\overline{D}(h, k, x) = 0$ at k_*–almost every point in E and $\overline{D}(h, k, x) = +\infty$ at h^*–almost every point in E.*

Proof. We suppose that $h \perp k$ on E. If it is not true that $\overline{\mathrm{D}}(h, k, x) = 0$ at k_*–almost every point in E then there is a subset E_0 of E and numbers $0 < c_1 < 1$ and $0 < c_2 < 1$ so that $k_*(E_0) > c_1$ and $\overline{\mathrm{D}}(h, k, x) > c_2$ everywhere on E_0. We may choose a full covering relation β_1 on E_0 in such a way that any packing π may be decomposed into $\pi = \pi_1 \cup \pi_2$ in such a way that

$$\sum_{(I,x) \in \pi_2} |h(I, x)| < c_1 c_2 / 2$$

and

$$\sum_{(I,x) \in \pi_1} |k(I, x)| < c_1 / 2.$$

Let

$$\beta_2 = \{(I, x) : |h(I, x)| > c_2 |k(I, x)|\}.$$

We compute $\mathrm{Var}(k, \beta_1 \cap \beta_2)$: if π is a packing from $\beta_1 \cap \beta_2$ then, using the above promised decomposition, we obtain

$$\sum_{(I,x) \in \pi} |k(I, x)| \leq \sum_{(I,x) \in \pi_1} |k(I, x)| + \sum_{(I,x) \in \pi_2} |h(I, x)| / c_2 < c_1.$$

As this holds for all packings and as $\beta_1 \cap \beta_2$ is evidently a fine covering relation on E_0 we obtain $k_*(E_0) \leq \mathrm{Var}(k, \beta_1 \cap \beta_2) \leq c_1$ which is a contradiction and provides the proof of the first assertion. Nearly identical proofs provide the other three assertions.

Dual versions of Theorems 5.16 and 5.15 are proved in the same manner.

THEOREM 5.17 *Let h and k be interval-point functions. Suppose that h^* is σ–finite on a set E of real numbers and that at h_*–almost every point $x \in E$ the derivative $\overline{\mathrm{D}}(h, k, x) = +\infty$. Then the set E may be decomposed into two disjoint subsets E_1 and E_2 so that $h_*(E_1) = k_*(E_2) = 0$ and consequently $h \perp k$ weakly on E.*

THEOREM 5.18 *If $h \perp k$ weakly on a set E then $\underline{\mathrm{D}}(h, k, x) = 0$ at k_*– almost every point in E and $\overline{\mathrm{D}}(h, k, x) = +\infty$ at h_*–almost every point in E.*

5.6   Characterization of singularity

We characterize now the notions of singularity in terms of derivation measure properties.

THEOREM 5.19 *Suppose that h and k are interval-point functions for which the measure h^* is σ–finite on a set E. Then the following are equivalent:*

1. *$h \perp k$ weakly on E.*

2. *The set E may be decomposed into two disjoint subsets E_1 and E_2 so that $h_*(E_1) = k_*(E_2) = 0$.*

3. *The derivative $\overline{D}(h, k, x) = +\infty$ at h_*–almost every point $x \in E$.*

This summarizes what we have just proved in the preceding sections and requires no proof. The theorem dual to Theorem 5.19 has also been proved.

THEOREM 5.20 *Suppose that h and k are interval-point functions for which the measure h^* is σ–finite on a set E. Then the following are equivalent:*

1. *$h \perp k$ on E.*

2. *The set E may be decomposed into two disjoint subsets E_1 and E_2 so that $h^*(E_1) = k^*(E_2) = 0$.*

3. *The derivative $\underline{D}(h, k, x) = +\infty$ at h^*–almost every point $x \in E$.*

For interval functions Theorem 5.19 can be somewhat improved.

THEOREM 5.21 *Suppose that h and k are interval functions for which the measure h^* is finite on a Borel set E. Then the following are equivalent:*

1. *$h \perp k$ weakly on E.*

2. *The set E may be decomposed into two disjoint Borel subsets E_1 and E_2 so that $h_*(E_1) = k_*(E_2) = 0$.*

3. *The derivative $\overline{D}(h, k, x) = +\infty$ at h_*–almost every point $x \in E$.*

4. *for $\delta > 0$ there are nonoverlapping intervals $I_1, I_2, \ldots I_m$ so that each* $|I_k| < \delta$, $\sum_{k=1}^{m} |k(I_k)| < \delta$ *and* $h_*(E \setminus \bigcup_{k=1}^{m} I_k) < \delta$.

Proof. The first three assertions are contained in the preceding theorem along with an obvious observation on the Borel structure of the set of points at which $\overline{D}(h, k, x) = +\infty$. To obtain the fourth assertion let us suppose that $h_*(E_1) = k_*(E_2) = 0$. Choose a fine covering relation β on E_2 so that $\text{Var}(k, \beta) < \delta$ and so that for every pair $(I, x) \in \beta$, $|I| < \delta$. Then by the Vitali theorem there is a finite packing $\pi \subset \beta$ so that

$$h_*(E_2 \setminus \bigcup_{(I,x) \in \pi} I) < \delta.$$

Thus, since $h_*(E_1) = 0$,

$$h_*(E \setminus \bigcup_{(I,x) \in \pi} I) < \delta$$

and certainly $\sum_{(I,x) \in \pi} |k(I)| < \delta$.

The other direction uses a standard measure-theoretic argument (cf. [16, p. 18]). Using the statement in 5.21(4) we choose intervals $I_{i1}, I_{i2}, I_{i2}, \ldots I_{im_i}$ so that each $|I_{ik}| < 2^{-i}$, so that

$$\sum_{k=1}^{m_i} |k(I_{ik})| < 2^{-i}$$

and

$$h_*(E \setminus \bigcup_{k=1}^{m_i} I_{ik}) < 2^{-i}.$$

We write

$$A = \bigcap_{i=1}^{\infty} \bigcup_{j=i+1}^{\infty} \bigcup_{k=1}^{m_j} I_{ik}^0$$

and obtain a Borel set for which we shall prove that $k_*(A) = 0$ and $h_*(E \setminus A) = 0$ which will establish assertion 5.21(2). The collection β_n of all pairs (I_{ik}, x) for $i > n$, $1 \le k \le m_i$, and x an interior point of I_{ik} is a fine covering relation on the set A and $\text{Var}(k, \beta_n) \le 2^{-n}$. This gives $k_*(A) = 0$.

To see that $h_*(E \setminus A) = 0$ note first that, for each $i = 1, 2, \ldots$ and $s > i$,

$$E \setminus \bigcup_{j=i+1}^{\infty} \bigcup_{k=1}^{m_j} I_{ik}^0 \subset E \setminus \bigcup_{k=1}^{m_s} I_{sk}^0.$$

Thus, since

$$h_*(E \setminus \bigcup_{k=1}^{m_s} I_{sk}) < 2^{-s},$$

we have

$$h_*(E \setminus \bigcup_{j=i+1}^{\infty} \bigcup_{k=1}^{m_j} I_{ik}^0) = 0.$$

Finally then, since the sequence of sets

$$E \setminus \bigcup_{j=i+1}^{\infty} \bigcup_{k=1}^{m_j} I_{ik}^0 \quad (i = 1, 2, \ldots)$$

covers $E \setminus A$, we have $h_*(E \setminus A) = 0$ as required.

5.7 Derivation properties and absolute continuity

In a similar manner to the material in the preceding sections information on the derivative f' can be used to claim absolute continuity properties; for example a function that has a bounded derivative is necessarily absolutely continuous. Theorem 5.22 gives a general version of this.

THEOREM 5.22 *Suppose that h and k are interval-point functions and that the derivative $\overline{D}(h, k, x) \leq M < \infty$ at h^*-almost every point x in a set E. Then $h \ll k$ on E.*

Proof. Let E_1 denote the set of points in E at which the derivate does not exceed M; by hypothesis $h^*(E \setminus E_1) = 0$ and so certainly $h \ll k$ on $E \setminus E_1$. It remains only to show that $h \ll k$ on E_1. The collection

$$\beta = \{(I, x) : |h(I, x)/k(I, x)| < M + 1\}.$$

is a full covering relation on E_1. If $\epsilon > 0$ and $\delta = \epsilon/(M + 1)$ then for any packing $\pi \subset \beta$ with

$$\sum_{\pi} |k(I, x)| < \delta$$

it is easy to check that

$$\sum_{\pi} |h(I, x)| < (M + 1) \sum_{\pi} |k(I, x)| < \epsilon$$

which establishes the required relation $h \ll k$ just on the set E_1. Together with the first result this proves that $h \ll k$ on E as required to complete the proof.

Note that because of Theorem 5.22 if the limit

$$\limsup_{I \to x} |h(I, x)/k(I, x)| < +\infty$$

holds at k^*-almost every point x in a set E, then, under appropriate measurability conditions, the relation $h \ll k$ on the set E may be established by applying Theorem 5.7 to the sets

$$E_n = \left\{ x \in E : \lim_{I \to x} |h(I, x)/k(I, x)| < n \right\}$$

for which we have the relation $h \ll k$ on each E_i. In particular we have by this method the following theorem.

THEOREM 5.23 *Suppose that h and k are interval functions, that the derivative $\overline{D}(h, k, x) < \infty$ at h^*-almost every point x in a Borel set E and that $h^*(E) < +\infty$. Then $h \ll k$ on E.*

5.8 Characterization of absolute continuity

In this section we provide a number of characterizations of the notion of absolute continuity in terms of the measure and derivation properties.

THEOREM 5.24 *Suppose that h and k are interval functions for which the measure h^* is finite on a Borel set E and k is weakly continuous. Then the following are equivalent:*

1. *The set E permits a Borel partition $E = E_1 \cup E_2$ so that $h_*(E_1) = 0$ and $h \ll k$ on E_2.*

2. *The set E has a Borel partition $E = \bigcup_{i=0}^{\infty} E_i$ so that $h_*(E_0) = 0$ and h is uniformly Lip(k) on each set E_k ($k \geq 1$).*

3. *The derivative $\overline{D}(h, k, x) < +\infty$ at h_*-almost every point $x \in E$.*

4. *$h_*(E \cap N) = 0$ for every Borel set N with $k_*(N) = 0$.*

5. *For all Borel subsets B of E,*

$$h_*(B) \leq h_*(B \cap E_0) + \int_{B \cap E_+} \overline{D}(h, k, x) \, dk_*(x) \leq h^*(B)$$

where E_0 and E_+ denote respectively the sets of points in E at which $\overline{D}(h, k, x)$ is zero and finite.

Proof. We show that (5) implies (4): the representation in (5) shows that $h_*(E \cap N) \leq h_*(E_0 \cap N)$ for every Borel set N with $k_*(N) = 0$. Then Theorem 4.6 shows that $h_*(E_0 \cap N) = 0$; this proves (4).

We show that (4) implies (3): the finiteness of the measure h^* on E and Theorem 4.5 show that

$$k_*(\{x \in E : \overline{D}(h, k, x) = +\infty\}) = 0.$$

Assuming (4) we then have that

$$h_*(\{x \in E : \overline{D}(h, k, x) = +\infty\}) = 0.$$

This proves (3).

We show that (3) implies (2): if (3) holds then the set

$$\{x \in E : \overline{D}(h, k, x) = +\infty\}$$

has zero h_*–measure and the set

$$\{x \in E : \overline{D}(h, k, x) < +\infty\}$$

has exactly the required partition because of Theorem 4.12 yielding (2).

We show that (2) implies (1): the representation in (2) shows that $h \ll k$ on each member of the sequence E_k for $k \geq 1$ and so Theorem 5.7 shows that $h \ll k$ on $\bigcup_{k=1}^{\infty} E_k$. This together with $h_*(E_0) = 0$ yields (1).

That (1) implies (4) follows immediately from the discussion in Section 5.3. Finally the proof is completed by showing that (3) implies (5). Let E_0, E_+ and E_∞ denote the sets of points in E at which the derivate $\overline{D}(h, k, x)$ is 0, positive and infinite respectively. We have from (3) that $h_*(E_\infty) = 0$ and we have from Theorems 4.7 and 3.23 that, for all Borel subsets B of E,

$$h_*(B \cap E_+) \leq \int_{B \cap E_+} \overline{D}(h, k, x) \, dk_*(x) \leq h^*(B \cap E_+).$$

Together these yield (1) and the proof is complete.

Finally we have the dual version of Theorem 5.24 with a somewhat similar proof.

THEOREM 5.25 *Suppose that h and k are interval functions for which the measure h^* is finite on a Borel set E and k is continuous. Then the following are equivalent:*

1. *$h \ll k$ weakly on E.*

2. *The derivative $\underline{D}(h, k, x) < +\infty$ at h_*–almost every point $x \in E$.*

3. *$h_*(E \cap N) = 0$ for every set Borel N with $k^*(N) = 0$.*

4. *For all Borel subsets B of E*

$$h_*(B) \leq h_*(B \cap E_0) + \int_{B \cap E_+} \underline{D}(h, k, x) \, dk^*(x) \leq h^*(B)$$

   where E_0 and E_+ denote respectively the sets of points in E at which $\underline{D}(h, k, x)$ is zero and finite.

Proof.   This is mostly similar to the proof of Theorem 5.24 and will be omitted.

6   Measures

6.1   Lebesgue measure

The Lebesgue measure and integral, playing as they do a central role in modern analysis, can be developed in many ways. Even a survey of the variety of approaches would be lengthy. There have long been controversies over which approach should first be learned; none of this much matters since all of the machinery of measure theory must be mastered in the end. Perhaps the last word on the subject should go to Temple [20] who is inspired by this situation to quote Kipling:

> There are nine and sixty ways of constructing tribal lays,
> And-every-single-one-of-them-is-right.

The material developed in the preceding sections adds two more ways to construct the Lebesgue measure and the Lebesgue integral. The characterizations here are particularly useful in connecting the derivation properties of functions with measure-theoretic properties.

If we write $\lambda(E) = |E|$ for the Lebesgue outer measure of a set E and let ℓ denote the interval function

$$\ell : I \to |I|$$

(which is the same as the interval function Δi for $i(x) = x$) then we shall have the following representations for the Lebesgue measure and integral:

$$|E| = \ell^*(E) = \ell_*(E)$$

and

$$(L) \int_E |f(t)| \, dt = (f\ell)^*(E) = (f\ell)_*(E).$$

Note that because of Theorem 3.27 the two identities $\ell^*(E) = \ell_*(E)$ and $(f\ell)^*(E) = (f\ell)_*(E)$ are really density assertions. Accordingly when they are invoked it is the Lebesgue density theorem that is implicitly being applied. This allows formal simplifications in a number of proofs.

THEOREM 6.1 *The measures ℓ^* and ℓ_* are precisely the Lebesgue outer measure.*

Proof.   For any open set G it is easy to see that $\ell^*(G) \le |G|$ and so the inequality $\ell^* \le \lambda$ must hold in general. The inequality $\ell_* \ge \lambda$ is a consequence of the Vitali theorem: any fine covering β of a bounded set E must contain a packing π with

$$\left| E \setminus \bigcup_{(I,x) \in \pi} I \right| = 0$$

and so

$$|E| \le \sum_{(I,x) \in \pi} |I|$$

and this proves that $|E| \le \mathrm{Var}(\ell, \beta)$. Since β is an arbitrary fine covering of E we may obtain the inequality $\lambda \le \ell_*$ in general. The two inequalities

$\ell^* \le \lambda$ and $\lambda \le \ell_*$ together with the obvious inequality $\ell_* \le \ell^*$ prove the statement of the theorem.

As remarked above the identity $\ell_* = \ell^*$ together with the density material in Section 3.10 can be used to obtain the familiar Lebesgue density theorem. We state this as a corollary.

COROLLARY 6.2 *Let E be a Lebesgue measurable set. Then*

$$\lim_{I \to x} \frac{|E \cap I|}{|I|} = \chi_E(x)$$

almost everywhere.

A consequence of Theorem 6.1 is to provide the following representations of the Lebesgue integral as full and fine upper integrals. This follows from Theorem 6.1 together with the material in Section 3.7.

COROLLARY 6.3 *Let f be a nonnegative, Lebesgue measurable function and E a Lebesgue measurable set E. Then*

$$(L) \int_E f(x)\,dx = V^*(f\ell, E) = V_*(f\ell, E).$$

6.2   Total variation measures

These same considerations apply to the study of the Lebesgue-Stieltjes measures. For any real function f we let Δf denote its associated interval function (thus $\Delta f([a,b]) = f(b) - f(a)$ for any interval $[a,b]$) and for continuous monotonic f let λ_f denote its usual Lebesgue-Stieltjes measure.

The two measures Δf^* and Δf_* will be called the *full and fine variational measures* associated with f. In the case of a continuous monotonic function f it will be seen that these measures are just the Lebesgue-Stieltjes measure λ_f associated with f; generally the utility of the measures goes further than this, but is limited to the study of functions that are at least VBG$_*$ on a set (see Section 7.2 for the terminology). Evidently if f is continuous then $\Delta f^*((a,b))$ is equal to the total variation of f on $[a,b]$. It again follows, as in the proof of Theorem 6.1, by an easy application of the Vitali covering theorem in the version for Lebesgue-Stieltjes measures that the measures have the following properties if f is continuous and monotonic nondecreasing:

1. $\Delta f^* = \Delta f_* = \lambda_f$.

2. Δf^* is $\mathcal{G}_\delta$ regular.

3. $\Delta f^*((a,b)) = \Delta f^*([a,b]) = f(b) - f(a)$.

This identity of Δf^* and Δf_* has many applications in the differentiation theory of monotonic functions. Note that this too, by Theorem 3.27, may be viewed as a reformulation of a density property.

6.3 Hausdorff measure

The measure theory developed here can be used to study various classically defined measures. We have seen that the Lebesgue and Lebesgue-Stieltjes measures on the real line may be characterized in the form h^* and h_* for an appropriate interval function. In a similar manner measures related to the Hausdorff s–dimensional measures may be investigated. Let us use λ^s to denote the Hausdorff s–dimensional measure. This can be defined as a limit of the outer measures, $\lambda^s = \lim_{\delta \to 0+} \lambda^s_\delta$, where λ^s_δ is defined by writing for any set E of real numbers

$$\lambda^s_\delta(E) = \inf \sum_{i=1}^\infty |I_i|^s$$

and the infimum is taken over all sequences $\{I_i\}$ of intervals of length less than δ covering the set E. For further definitions and properties any number of texts will suffice; the best references are probably Rogers [15] and Falconer [4]. From these we require only the rudiments of the theory together with the fundamental density properties of the measures.

Recall that we use ℓ to denote the interval function $I \to |I|$ that assigns the length of the interval I to each closed interval I. Then the measures μ^s and μ_s are defined for any $0 < s \le 1$ as

$$\mu^s(E) = V^*(\ell^s, E)$$

and

$$\mu_s(E) = V_*(\ell^s, E).$$

The case $s < 1$ is quite different from the case $s = 1$ for the measures μ^s and μ_s. We have already seen that $\mu^1 = \mu_1$; for $s < 1$ the two measures are never the same (except for the trivial values 0 and $+\infty$). This is proved in Theorem 6.5 below; note here that the measure μ^s, since it assumes only the values 0 and $+\infty$, makes all sets of real numbers μ^s–measurable. Even so one can define a "full dimension" using μ^s that can be seen to be greater than the Hausdorff dimension. The "fine dimension" defined using the measure μ_s is precisely, because of Theorem 6.4 which we now prove, the usual Hausdorff dimension.

THEOREM 6.4 *The measure μ_s is equivalent to the s–dimensional Hausdorff measure in the sense that for any λ^s–measurable set E the identity $\lambda^s(E) = \mu_s(E)$ holds.*

Proof. This follows from well-known properties of the the s–dimensional Hausdorff measure. To obtain $\lambda^s(E) \leq \mu_s(E)$ we choose an arbitrary fine covering relation β on E. By the Vitali covering theorem for λ^s (see Falconer [4, Theorem 1.10, p. 11]) there is a packing $\pi \subset \beta$ with

$$\lambda^s(E) \leq \sum_{(I,x)\in\pi} |I|^s + \epsilon$$

if $\lambda^s(E) < \infty$ or so that $\sum_{(I,x)\in\pi} |I|^s = +\infty$ if $\lambda^s(E) = \infty$. In either case

$$\lambda^s(E) \leq \text{Var}(\ell^s, \beta) + \epsilon.$$

Since β is an arbitrary fine covering relation on E and ϵ an arbitrary positive number we must have $\lambda^s(E) \leq \mu_s(E)$.

We obtain a supplementary result now: each set E that has $\lambda^s(E)$–measure zero must have $\mu_s(E)$–measure zero. Suppose that $\lambda^s(E) = 0$ and that $\epsilon > 0$. For each $n = 1, 2, 3, \ldots$ there is a covering of the set E by a sequence of open intervals $\{I_{nj}^0\}$ each of length less than $1/n$ and such that $\sum_{j=1}^{\infty} |I_{jn}|^s < \epsilon/2^{n+1}$. The family β of all pairs (I_{nj}, x) for $x \in E$ and for such intervals is a fine covering relation on E and certainly

$$\mu_s(E) \leq \text{Var}(\ell^s, \beta) < \epsilon.$$

Thus $\mu_s(E) = 0$ as we require.

For the inequality $\mu_s(E) \leq \lambda^s(E)$ we need to investigate the density

$$d^s(E, x) = \limsup \frac{\lambda^s(E \cap I)}{|I|^s}$$

taken as $|I| \to 0$ with $x \in I^0$. For any λ^s–measurable set E with finite λ^s–measure it is known (see Falconer [4, Theorem 2.3,p. 24]) that $d^s(E, x) \geq 1$ at λ^s–almost every point of E; in view of the previous paragraph, this holds too at μ_s–almost every point of E. Let $0 < \alpha < 1$. The collection β of all pairs (I, x) with $x \in E \cap I^0$ and

$$\lambda^s(E \cap I) > \alpha |I|^s$$

is a fine covering relation on a set $E_0 \subset E$ where $\mu_s(E \setminus E_0) = 0$. But for any packing π from β

$$\sum_{(I,x) \in \pi} |I|^s \leq \sum_{(I,x) \in \pi} \alpha^{-1} \lambda^s(E \cap I) \leq \alpha^{-1} \lambda^s(E)$$

and so

$$\mu_s(E_0) \leq \text{Var}(\ell^s, \beta) \leq \alpha^{-1} \lambda^s(E).$$

Let $\alpha \to 1$ and use $\mu_s(E \setminus E_0) = 0$, to obtain that $\mu_s(E) \leq \lambda^s(E)$ as required to complete the proof.

We turn now to the larger of the two measures μ^s and μ_s showing that this measure is essentially trivial. This has appeared in the literature as the "packing measure" of Tricot (see [23], [18] and [19]) although differently defined.

THEOREM 6.5 *Let $0 < s < 1$ and let E be a set of real numbers. Then $\mu^s(E)$ is either 0 or $+\infty$.*

Proof. We show that any bounded set E for which $\mu^s(E) < +\infty$ is contained in a Borel set of type $\mathcal{F}_\sigma$ that has μ^s –measure zero. Suppose then that β is a full covering relation on the set E with $\text{Var}(\ell^s, \beta) < +\infty$. By the covering Lemma 2.11 this permits us to choose an increasing sequence of sets $\{E_n\}$ whose union is E such that any pair (I, x) with $|I| < 1/n$, $x \in E_n \cap I^0$ must be in β.

We fix n, write $F_n = \overline{E_n}$ and show that $\mu^s(F_n) = 0$. Define β_1 as the collection of all pairs (I, x) with $|I| < 1/n$ and $x \in F_n \cap I^0$. Note that for any pair (I, x) in β_1 there is a corresponding pair (I, z) in β.

Certainly β_1 must be a (uniform) full covering relation on F_n so that $\mu^s(F_n) \leq \text{Var}(\ell^s, \beta_1) \leq \text{Var}(\ell^s, \beta) < +\infty$. Since s is less than 1 this requires

F_n to have Lebesgue measure zero. To see this observe that if $\delta > 0$ and $c < 1$ then for any full covering relation γ on F_n chosen so that $\gamma \subset \beta_1$ and each interval-point pair $(I, x) \in \gamma$ has $|I| < \delta$ we will have $\mathrm{Var}(\ell, F_n) > c|F_n|$. Thus

$$\mathrm{Var}(\ell^s, \gamma) \geq \delta^{s-1} \mathrm{Var}(\ell, \gamma) \geq \delta^{s-1} c |F_n|.$$

From this it follows that

$$\mathrm{Var}(\ell^s, \beta) \geq \delta^{s-1} |F_n|$$

for all $\delta > 0$ which is possible only if $|F_n| = 0$.

Let $\{I_k\}$ denote the sequence of intervals contiguous to F_n in some bounding interval $[a, b]$. We claim that $\sum_{k=1}^{\infty} |I_k|^s < +\infty$. If not then from this sequence we may extract a finite collection $\{J_1, J_2, \ldots, J_N\}$ so that each interval $|J_k| < 1/n$ and

$$\sum_{i=1}^{N} |J_i|^s \geq 4\mathrm{Var}(\ell^s, \beta).$$

But each of these intervals J_i may be enlarged to an interval J_i' of length less than $1/n$ that contains a point of E_n as an interior point and so that no more than any three of the new intervals overlap. These pairs (J_i', z_i) for appropriate points z_i belong to β and that requires

$$\sum_{i=1}^{N} |J_i|^s \leq \sum_{i=1}^{N} |J_i'|^s \leq 3\mathrm{Var}(\ell^s, \beta)$$

which is impossible.

We now compute $\mu^s(F_n)$. Let C denote the set of points in F_n that are isolated on either side and let $\epsilon > 0$. Choose N so large that

$$\sum_{i=N}^{\infty} |I_i|^s < \epsilon/2$$

and select a full covering relation β_1 on $F_n \setminus C$ so that every pair $(I, x) \in \beta_1$ has $I \cap I_i = \emptyset$ for $i = 1, 2, \ldots, N - 1$. Consider any packing $\pi \subset \beta_1$: if $(I, x) \in \pi$ then

$$I \setminus \bigcup_{i=N}^{\infty} I_i = I \setminus \bigcup_{i=1}^{\infty} I_i \subset F_n.$$

Since F_n has Lebesgue measure zero we obtain

$$|I| \le \sum_{i=N}^{\infty} |I \cap I_i|$$

and hence

$$|I|^s \le \sum_{i=N}^{\infty} |I \cap I_i|^s.$$

But this gives

$$\mathrm{Var}(\ell^s, \beta_1) \le \sum_{(I,x) \in \pi} \sum_{i=N}^{\infty} |I \cap I_i|^s.$$

No interval I_k can intersect more than two of the intervals I from such a packing and this gives the upper estimate

$$\mu^s(F_n \setminus C) \le \mathrm{Var}(\ell^s, \beta_1) \le 2 \sum_{i=N}^{\infty} |I_i|^s < \epsilon.$$

Consequently $\mu^s(F_n \setminus C) = 0$; since C is countable $\mu^s(C) = 0$ too and we have that $\mu^s(F_n) = 0$ as required to complete the proof.

6.4 Density theorems

Estimates on the size of the Hausdorff measure $\lambda^s(E)$ of a set E are frequently difficult to obtain. One of the most useful tools in such a theory is a form of density theorem; see Taylor and Tricot [18, Theorem 2.1, p. 682]. For the measures μ^s and μ_s a version of this density theorem is available directly from the estimates in Section 4.3.

We shall abbreviate the derivate $\overline{D}(\Delta f, \ell^s, x)$ as $\overline{D}_s(f, x)$. Thus for any real function f we are writing

$$\overline{D}_s(f, x) = \limsup \frac{|f(y) - f(z)|}{|y - z|^s}$$

where the limit is taken as $y, z \to x$ with $y < x < z$. Similarly $\underline{D}_s(f, x)$ denotes the limit inferior of the same expression.

THEOREM 6.6 *Let f be a continuous real function and E a set of real numbers. Then*

$$\frac{\Delta f_*(E)}{\sup_{x\in E} \overline{D}_s(f,x)} \leq \mu_s(E) \leq \frac{\Delta f^*(E)}{\inf_{x\in E} \overline{D}_s(f,x)}$$

and

$$\frac{\Delta f_*(E)}{\sup_{x\in E} \underline{D}_s(f,x)} \leq \mu^s(E) \leq \frac{\Delta f^*(E)}{\inf_{x\in E} \underline{D}_s(f,x)}.$$

Proof. These estimates follow directly from Corollary 4.7.

We now present a density theorem for the measure μ_s; because of Theorem 6.5 there is no nontrivial density theorem for the measure μ^s. Note that in the statements of Theorem 6.7 and Theorem 6.8 the condition on the lower density holds almost everywhere with respect to the larger measure.

THEOREM 6.7 *Let E be μ_s-measurable and suppose that $\mu_s(E) < +\infty$ and $0 < s < 1$. Then*

$$\liminf_{I\to x} \frac{\mu_s(E \cap I)}{|I|^s} = 0$$

for μ^s-almost every $x \in E$ and

$$\limsup_{I\to x} \frac{\mu_s(E \cap I)}{|I|^s} = \chi_E(x)$$

for μ_s-almost every x.

Proof. For each integer n let E_n denote the set of points in E at which

$$\liminf_{I\to x} \frac{\mu_s(E \cap I)}{|I|^s} > 1/n.$$

Then the collection β of all pairs (I, x) for which

$$\mu_s(E \cap I) > n^{-1}|I|^s$$

is a full covering relation on E_n. Thus

$$\mu^s(E_n) \leq V^*(\ell^s, \beta) < n\mu_s(E) < +\infty.$$

Each $\mu^s(E_n)$ is finite; thus by Theorem 6.5 each must vanish and the first part of the theorem now follows easily. The second part follows directly from Theorem 3.26.

By the same methods we can prove the following differentiation result for any interval function.

THEOREM 6.8 *Let h be an interval function and suppose that h^* is σ-finite on E. Then*

$$\liminf_{I \to x} \frac{|h(I)|}{|I|^s} = 0$$

for μ^s–almost every $x \in E$.

6.5   Hausdorff dimension

The two families of measures $\{\mu^s : 0 \leq s \leq 1\}$ and $\{\mu_s : 0 \leq s \leq 1\}$ can be used in the standard manner to define two rarefaction indices. For any set E of real numbers we call the number

$$\inf \{s : \mu^s(E) = 0\} = \sup \{s : \mu_s(E) = +\infty\}$$

the *full dimension* of E and we call the number

$$\inf \{s : \mu_s(E) = 0\} = \sup \{s : \mu_s(E) = +\infty\}$$

the *fine dimension* of E. Certainly the latter is the familiar Hausdorff dimension in view of Theorem 6.4. The full dimension exceeds the fine dimension and for some sets they are distinct. The larger dimension has been introduced in this theory by Tricot [23] and studied in the articles [18] and [19] under the term "packing dimension" although the definition is different from ours. Note that other differentiation bases would give a similar pair of measures and so related scales of dimensions.

The clearest illustration of how this pair of dimensional ideas arises naturally in certain familiar situations is given in the computations of the dimension of symmetric perfect sets. We outline a result that is observed in [18] by different methods. If $\{\xi_n\}$ is a sequence of numbers from the interval $(0, 1/2)$ the *symmetric perfect set* $C(\{\xi_n\})$ is defined by a Cantor-like procedure (cf. [8]) as

$$C(\{\xi_n\}) = \bigcap_{n=1}^{\infty} \bigcup_{k=1}^{2^n} E_n{}^k$$

where $E_1{}^1 = [0, \xi_1]$, $E_1{}^2 = [1 - \xi_1, 1]$, $E_2{}^1 = [0, \xi_1\xi_2]$, $E_2{}^2 = [\xi_1 - \xi_1\xi_2, \xi_1]$, and so on in such a way that at the n–th stage there are 2^n intervals each of length $\xi_1\xi_2\xi_3 \ldots \xi_n$. As in [8, p. 14] we may call these 'black" intervals; the contiguous intervals, the "white" intervals, form a sequence so that at the n–th stage there are 2^{n-1} open intervals of length $\xi_1\xi_2\xi_3 \ldots \xi_{n-1}(1 - 2\xi_n)$. The dimensions for such sets are expressed in the following theorem.

THEOREM 6.9 *Let $C(\{\xi_n\})$ be a symmetric perfect set. Then the full and fine dimensions of $C(\{\xi_n\})$ are given by the expressions*

$$\limsup_{n\to\infty} \frac{n \log 2}{-\log(\xi_1\xi_2\xi_3 \ldots \xi_n)} \quad and \quad \liminf_{n\to\infty} \frac{n \log 2}{-\log(\xi_1\xi_2\xi_3 \ldots \xi_n)}.$$

Proof. The expression for the fine (Hausdorff) dimension seems to be well known and will not be presented here. For the full dimension let us suppose that

$$s_1 < \limsup_{n\to\infty} \frac{n \log 2}{-\log(\xi_1\xi_2\xi_3 \ldots \xi_n)} < s_2.$$

Write $\{I_k\}$ for the sequence of intervals contiguous to $C = C(\{\xi_n\})$ and write N for the set of endpoints of these intervals. We see that $\sum_{k=1}^{\infty} |I_k|^{s_2}$ converges while $\sum_{k=1}^{\infty} |I_k|^{s_1}$ diverges and it is this difference that allows the distinction.

Define the two interval functions h_1 and h_2 by writing

$$h_1(I) = \sup |I \cap I_k|^{s_1}$$

and

$$h_2(I) = \sum_{k=1}^{\infty} |I \cap I_k|^{s_2}.$$

We note easily that $h_2(I) \le |I|^{s_2}$ and we have $h_1(I) \ge |I|^{s_1}$; the latter is obvious while for the former we note that in the case where $C(\{\xi_n\})$ has measure zero (which is the only case we need consider) every interval I satisfies $|I| = \sum_{k=1}^{\infty} |I \cap I_k|$ and hence $|I|^{s_2} \le \sum_{k=1}^{\infty} |I \cap I_k|^{s_2}$. Evidently then $\mu^{s_1} \ge h_1{}^*$ and $\mu^{s_2} \le h_2{}^*$.

We show that $h_2{}^*(C)$ is finite and that $h_1{}^*(C)$ is infinite; from this the estimate in the theorem for the full dimension must follow. Since h_1 and h_2 are continuous $h_1{}^*(N) = h_2{}^*(N) = 0$ and so we may consider just the computations of $h_1{}^*(C \setminus N)$ and $h_2{}^*(C \setminus N)$. That $h_2{}^*(C \setminus N)$ is finite follows easily from the convergence of the series $\sum_{k=1}^{\infty} |I_k|^{s_2}$.

Suppose now, contrary to what we wish to prove, that $h_1{}^*(C \setminus N)$ is finite. Choose a full covering relation β on C so that $\mathrm{Var}(h_1, \beta) < \infty$. Then by Theorem 2.11 and the Baire category theorem there is a portion $C \cap (a, b)$ of C so that for every interval-point pair (I, x) with $I \subset (a, b)$ and $x \in I^0 \cap C$ there is a pair (I, z) that belongs to β. As $\sum_{k=1}^{\infty} |I_k|^{s_1}$ diverges and from the self-similarity of the set C it follows that there are $I_{k_1}, I_{k_2}, I_{k_3}, \dots, I_{k_m}$ subintervals of $[a, b]$ so that the sum $\sum_{p=1}^{m} h(I_{k_p})$ exceeds $\mathrm{Var}(h_1, \beta)$. But this yields a contradiction since we may enlarge each interval I_{k_p} slightly to an interval J_{k_p} in such a way that the resulting intervals remain disjoint and now each include a point of C (say x_{k_p}) in their interior. The collection $\{(J_{k_p}, z_{k_p})\}$ for some choice of z_{k_p} is contained in β and so

$$\sum_{p=1}^{m} h(I_{k_p}) \leq \sum_{p=1}^{m} h(J_{k_p}) \leq \mathrm{Var}(h_1, \beta).$$

This contradiction shows that $h_1{}^*(C)$ is infinite and the proof is complete.

7   Real functions

We turn now to a study of the properties of continuous real functions. All the results arise by considering the derivation and variation properties of the interval function Δf for a continuous real function f. We have restricted attention throughout to continuous functions because of the particular differentiation basis studied here; a function f may be nonconstant and yet have everywhere a derivative (in the present sense) which vanishes. This peculiarity is easily avoided by considering just continuous functions; we wish to give statements that are true for the ordinary derivative and we can achieve this merely by restricting attention to continuous functions.

7.1   Monotonic functions

The classical differentiation and integration theory for the class of monotonic functions can be developed in a number of ways. As an illustration of the techniques here we show how this material can arise from the general study of interval functions given thus far. Note that the key tool is the Vitali covering theorem (as is the case in most presentations) but the technical applications are transparent and economical, given the machinery that has been developed. In particular it is only the measure identity $\Delta f^* = \Delta f_*$ that needs to be used in order to obtain many of the deeper differentiation properties of these functions.

The usual notions of bounded variation, absolute continuity and singularity of functions may be used; in addition we extend these by using the language of Section 5. We say f is *absolutely continuous* on a set E if $\Delta f \ll \ell$ on E and that f is *singular* on E if $\Delta f \perp \ell$ on E. Functions f_1 and f_2 are *mutually singular* on a set E if $\Delta f_1 \perp \Delta f_2$ on that set. The phrase "almost everywhere" without mention of a measure refers, as usual, to the Lebesgue measure and we may use either of the representations ℓ^* or ℓ_*.

Our first classical result, due to Lebesgue, concerns the differentiability properties of monotonic functions. The proof is mostly based on a single computational idea: if the derivates of the function f are finite and nonzero on a set E then the "spread" between them can be exhibited as

$$\int_E \left(\overline{D} f(x) - \underline{D} f(x) \right) dx \leq \Delta f^*(E) - \Delta f_*(E). \qquad (31)$$

Together with the density identity $\Delta f^* = \Delta f_*$ this shows rather transparently why the derivative must exist outside of a set small in the two senses of the theorem. The proof is elementary except for the appeal to this density identity.

THEOREM 7.1 (Lebesgue differentiation theorem) *If f is continuous and nondecreasing then $f'(x)$ exists as a finite real number almost everywhere and $f'(x)$ exists positive or infinite Δf^*-almost everywhere.*

Proof. The upper and lower derivates $\overline{D} f(x)$ and $\underline{D} f(x)$ of f are exactly the derivates $\overline{D}(\Delta f, \ell, x)$ and $\underline{D}(\Delta f, \ell, x)$ in the language of Section 4 and so we may appeal to the theory developed there. The proof is given by an analysis of the measures of the five sets we now define:

$$
\begin{aligned}
N_1 &= \{x : \overline{D} f(x) = +\infty\}, \\
N_2 &= \{x : \overline{D} f(x) = 0\}, \\
N_3 &= \{x : 0 = \underline{D} f(x) < \overline{D} f(x)\}, \\
N_4 &= \{x : 0 < \underline{D} f(x) < \overline{D} f(x) < +\infty\}, \\
N_5 &= \{x : \underline{D} f(x) < \overline{D} f(x) = +\infty\}.
\end{aligned}
$$

That $|N_1| = 0$ follows from Corollary 4.5 since the measure Δf^* is obviously finite on every bounded set. We obtain $\Delta f^*(N_2) = 0$ and $\Delta f_*(N_3) = 0$ directly from Corollary 4.6. Since Δf^* and Δf_* are identical this gives $\Delta f^*(N_2) = \Delta f^*(N_3) = 0$. But the vanishing of $\Delta f^*(N_3)$ requires $\overline{D} f(x)$ to vanish almost everywhere on N_3 and so $|N_3| = 0$ too.

The set N_4 is handled immediately by the inequality (31) cited before the statement of the theorem; this follows from the inequalities of Theorem 4.7. combined with our representation of the Lebesgue integral in Theorem 6.3. From this we obtain $|N_4| = 0$. The same source provides the inequality

$$\Delta f_*(N_4) \leq \int_{N_4} \overline{D} f(x) \, dx$$

and so $\Delta f_*(N_4) = 0$ too. Again the identity of Δf^* and Δf_* supplies that $\Delta f^*(N_4) = 0$ too.

Finally, again from the same source, we have the inequality

$$\Delta f_*(N_5) \leq (L) \int_{N_5} \underline{D} f(x) \, dx.$$

But as $N_5 \subset N_1$ and N_1 has measure zero this means that $\Delta f_*(N_5) = 0$. Yet again the identity of Δf^* and Δf_* supplies that $\Delta f^*(N_5) = 0$ too. These measure estimates complete the proof.

The well known decomposition theorem of Jordan follows almost immediately from the elementary decomposition

$$\Delta f = (\Delta f)^+ - (\Delta f)^-$$

where $(\Delta f)^+(I) = \max\{\Delta f(I), 0\}$ and $(\Delta f)^-(I) = \max\{-\Delta f(I), 0\}$. In the version here the differentiation properties of the decomposition are provided too.

THEOREM 7.2 (Jordan decomposition) *Let f be a continuous function that has bounded variation in every compact interval. Then there are nondecreasing, continuous functions g_1 and g_2 so that*

1. $f = g_1 - g_2$.

2. g_1 and g_2 are mutually singular.

3. $T = g_1 + g_2$ is the total variation of f.

4. $(\Delta f)^+ \equiv \Delta g_1$, $(\Delta f)^- \equiv \Delta g_2$ and $|\Delta f| \equiv \Delta T$.

5. For any Borel set E

$$\Delta f^*(E) = \Delta f_*(E) = \Delta T^*(E) = \Delta g_1{}^*(E) + \Delta g_2{}^*(E).$$

6. At almost every point x the following identities for the derivatives hold:

(a) $f'(x) = g_1'(x) - g_2'(x)$.

(b) $g_1'(x)g_2'(x) = 0$.

(c) $f'(x) = g_1'(x)$ or $f'(x) = g_2'(x)$.

(d) $(|\Delta f|)'(x) = T'(x) = |f'(x)| = g_1'(x) + g_2'(x)$.

(e) $(\Delta f^+)'(x) = g_1'(x)$ and $(\Delta f^-)'(x) = g_2'(x)$.

Proof.   Let g_1 and g_2 denote functions associated with the total variation functions for $(\Delta f)^+$ and $(\Delta f)^-$ (these are unique up to an additive constant). These latter two interval functions are subadditive and have bounded variation so that Theorem 3.6 may be used and then Theorem 7.1 applied to show the almost everywhere differentiability. The elementary identities $\Delta f = (\Delta f)^+ - (\Delta f)^-$, $\Delta g_1 \equiv (\Delta f)^+$, $\Delta g_2 \equiv (\Delta f)^-$ and $(\Delta f)^+(\Delta f)^- = 0$ together with the identity $\Delta f^*(E) = \Delta f_*(E)$ (see Section 6.2), the material on exact derivatives (see Section 4.5) and Theorem 3.25 then supply the rest of the theorem.

Continuing our catalogue of classical applications we have the decomposition of monotonic functions into their absolutely continuous and singular parts; as in the standard proofs the focus is on the set of points of differentiability.

THEOREM 7.3 (Lebesgue decomposition) *A continuous, nondecreasing function f may be written in the form*

$$f(t) = f_1(t) + \sum_{i=1}^{\infty} g_k(t)$$

where f_1 is a continuous, nondecreasing singular function and each g_k is nondecreasing and Lipschitz. If $f_2 = \sum_{i=1}^{\infty} g_k$ then f_2 is absolutely continuous and f_1 and f_2 are mutually singular.

Proof. Let E_1 denote the set of points x at which $f'(x)$ is zero, let E_2 denote the set of points x at which $f'(x)$ is finite and positive and let E_3 denote the set of points x at which $f'(x)$ is $+\infty$. By Theorem 7.1 we know that

$$\Delta f^*(R \setminus (E_1 \cup E_2 \cup E_3)) = 0$$

and, by Theorem 4.6, $\Delta f^*(E_1) = 0$.

Consequently for $t > 0$ (with a similar expression for $t < 0$)

$$f(t) - f(0) = \Delta f^*((0, t) \cap E_3) + \Delta f^*((0, t) \cap E_2)$$

expresses f as the sum of two functions f_1 and f_2.

The function $f_1(t) = \Delta f^*((0, t) \cap E_3)$ can be seen to be singular. We show that $\Delta f_1^*(\mathbb{R} \setminus E_3) = 0$; since E_3 has Lebesgue measure zero this shows that f_1 is singular. Let $\epsilon > 0$ and choose an open set G containing the complement of E_3 and such that $\Delta f^*(G \cap E_3) < \epsilon$ (see, for example, [15, Theorem 22, p. 35]). Then for any full covering relation β on $\mathbb{R} \setminus E_3$

$$\Delta f_1^*(\mathbb{R} \setminus E_3) \leq \mathrm{Var}(\Delta f_1, \beta(G)) < \epsilon.$$

Indeed if $\pi \subset \beta(G)$ is a packing then

$$\sum_{(I,x) \in \pi} \Delta f_1(I) = \sum_{(I,x) \in \pi} \Delta f^*(I \cap E_3) \leq \Delta f^*(G \cap E_3) < \epsilon.$$

Since ϵ is arbitrary we obtain $\Delta f_1^*(\mathbb{R} \setminus E_3) = 0$ as required.

The function $f_2(t) = \Delta f^*((0, t) \cap E_2)$ can be seen to be absolutely continuous and can be displayed in the series representation of the theorem using Theorem 4.11. The details are straightforward and are, in any case, duplicated in the more general proof of Theorem 7.18.

The next assertion is our version of the well known de la Vallée Poussin decomposition theorem.

THEOREM 7.4 (de la Vallée Poussin) *Suppose that f is a continuous and nondecreasing function and E a Borel set. Then $f'(x)$ exists a.e. and*

$$\Delta f^*(E) = (L) \int_E f'(x) \, dx + \Delta f^* \left(\{x \in E : f'(x) = +\infty\} \right).$$

Moreover

$$\Delta f^*(E) = (L) \int_E f'(x) \, dx$$

if and only if f is absolutely continuous on E.

Proof. The decomposition follows as in the proof of Theorem 7.3. For if E_1, E_2 and E_3 denote, respectively, the sets of points in E at which $f'(x)$ fails to exist (finite or infinite), at which it is finite and at which it is infinite then these are disjoint Borel sets (see Theorem 4.2) with union E and with $\Delta f^*(E_1) = 0$. The representation

$$\Delta f^*(E \cap E_2) = (L) \int_{E \cap E_2} f'(x)\, dx$$

follows from Theorem 4.7 and the fact (Theorem 6.3) that $(f'\ell)^*(E \cap E_2)$ may be expressed as the Lebesgue integral $(L) \int_{E \cap E_2} f'(x)\, dx$.

We complete our study of the structure of monotonic functions by characterizing the singular functions.

THEOREM 7.5 *Let f be a continuous, nondecreasing function and E a bounded Borel set. The following assertions are equivalent:*

1. *$\Delta f \perp \ell$ on E.*

2. *$\Delta f \perp \ell$ weakly on E.*

3. *$f'(x) = 0$ at almost every $x \in E$.*

4. *$f'(x) = +\infty$ at λ_f–almost every $x \in E$.*

5. *There is a subset $N \subset E$ such that $\lambda_f(E \setminus N) = |N| = 0$.*

6. *$\sqrt{\Delta f \ell} \equiv 0$ on E.*

7. *For every $\delta > 0$ there are nonoverlapping intervals $I_1, I_2, \ldots I_m$ so that $\sum_{k=1}^m |I_k| < \delta$ and $\lambda_f \left(E \setminus \bigcup_{k=1}^m I_k \right) < \delta$.*

8. *For every $\delta > 0$ there are nonoverlapping intervals $I_1, I_2, \ldots I_m$ so that each $|I_k| < \delta$, $\sum_{k=1}^m \Delta f(I_k) < \delta$ and $|E \setminus \bigcup_{k=1}^m I_k| < \delta$.*

The proof is obtained by a direct application of the general results in Section 5.6 and can be omitted. As a corollary we can obtain the following version of a result of Kober [9] (cf. also Mauldon [14]). If f is a homeomorphism then the singularity conditions on a set E map over easily to singularity conditions for f^{-1} on the image set.

COROLLARY 7.6 *Let f be a continuous, strictly increasing function and E a bounded Borel set. Then f is singular on E if and only if f^{-1} is singular on the set $f[E]$.*

7.2   Functions having σ–finite variation

If f is a continuous function having bounded variation then both the measures Δf^* and Δf_* agree with the usual Lebesgue-Stieltjes measure λ_v associated with the total variation function v of f. This follows from the material in Section 6.2 and Theorem 7.2. In general for an arbitrary continuous function f we shall consider the two measures Δf^* and Δf_* as together describing the total variation of f.

In the classical theory the differential structure of monotonic functions is carried over to functions of bounded variation by way of the Jordan decomposition theorem. This is still not sufficient to describe all situations in differentiation theory. For example a differentiable function f need not have bounded variation (unless f' is Lebesgue integrable). A further extension (due to Lusin and Denjoy) gives rise to the class of VBG$_*$ functions; Saks [17, Chapter VII] gives a full account. Briefly a function f is VB$_*$ on a set E if the sums

$$\sum_{i=1}^{n} \sup_{c,d \in [a_i,b_i]} |f(d) - f(c)|$$

remain bounded taken over all sequences $\{[a_i, b_i]\}$ of non-overlapping intervals with endpoints in E; a function f is VBG$_*$ on a set E if it is VB$_*$ on each member of a sequence of sets covering E.

There is an obvious analogy to the situation of finite and σ–finite measures. As we shall see in this section this can be viewed as more than an analogy by utilizing the measures Δf^* and Δf_*. In addition the VBG$_*$ concept will be viewed as a form of density theorem for these measures. In addition to these standard notions of VB$_*$ and VBG$_*$ we require some auxiliary notions: let us say that a function f has *uniformly bounded variation* on a set E if there are positive numbers M and δ so that the sums $\sum_{i=1}^{n} |\Delta f(I)|$ remain bounded by M for any sequence $\{I_1, I_2, \ldots, I_n\}$ of nonoverlapping intervals each of length less than δ and each containing a point of E as an interior point. This clearly extends the notion of bounded variation but in a manner different from the VB$_*$ notion.

Observe first that the measure Δf_* is invariably σ–finite. This is not the case for the measure Δf^*; in fact for everywhere nondifferentiable functions it can be shown that the smaller measure Δf_* vanishes while the full variation measure is non-σ–finite.

THEOREM 7.7 *Let f be a continuous function. Then the measure Δf_* is σ–finite.*

Proof. We define a number of sets. Let E_1 be the set of all points x for which there are sequences $x_n < x < y_n$ with $y_n - x_n \to 0$ and $f(x_n) = f(y_n)$ for all n; let E_2 be the set of all points x at which f has a strict extremum. Let E_3 (and E_4) be the sets of points x at which there is a $\delta(x) > 0$ so that $f(x) < f(y)$ (or $f(x) > f(y)$ for E_4) for all $x < y < x + \delta(x)$. Finally let E_5 and E_6 denote the analogous left hand versions of E_3 and E_4. This exhausts all possibilities. Our theorem is proved by showing that $\Delta f_*(E_1) = \Delta f_*(E_2) = 0$ and that Δf_* is σ–finite on each of the sets E_3, E_4, E_5 and E_6.

That $\Delta f_*(E_1) = 0$ follows easily from the fact that $\{(I, x) : \Delta f(I) = 0\}$ is a fine covering relation on E_1 and that $\Delta f_*(E_2) = 0$ follows from the continuity of f since this set is countable. Let us show that Δf_* is σ–finite on E_3. The function δ may be used as in Theorem 2.10 to construct a partition $\{A_n\}$ of the set E_3 in such a way that f is increasing on each member of the partition with the sets A_n bounded; let B_n denote the points in A_n that are not isolated on any side and construct the collection

$$\beta = \{([a, b], x) : a, b \in B_n, a < x < b\}.$$

Evidently this is a fine covering relation on B_n and $\mathrm{Var}(\Delta f, \beta)$ is finite. The set $A_n \setminus B_n$ is countable and so the measure $\Delta f_*(A_n) < \infty$ showing that Δf_* is σ–finite on E_3. In a similar way we may show that Δf_* is σ–finite on each of E_4, E_5 and E_6 and so the theorem is proved.

In contrast to the theorem just proved we show that the σ–finiteness of the full measure Δf^* has strong implications for the differentiation structure of the function f.

THEOREM 7.8 *Let f be a continuous function and let E be a Borel set of real numbers. Then the following assertions are equivalent:*

1. Δf^* is σ–finite on E.

2. E has a Borel partition $\bigcup_{i=1}^{+\infty} E_i$ such that f has uniformly bounded variation on each set $\overline{E}_i$.

3. E may be written as an increasing union $\bigcup_{i=1}^{+\infty} E_i$ of Borel sets in such a way that, for each n, $\Delta f \equiv \Delta g_n$ on E_n for some sequence $\{g_n\}$ of continuous functions, each of bounded variation.

4. f is VBG_* on E.

5. There is a continuous, increasing function g so that $\overline{D}(\Delta f, \Delta g, x)$ is finite at every point of E.

6. There is a homeomorphism g so that the composition $f \circ g$ has a finite derivative everywhere in $g^{-1}(E)$.

7. There is a continuous nondecreasing function g_1 and a function g_2 so that $\Delta f \equiv g_2 \Delta g_1$ on E.

8. $\Delta f^*(E \cap T) = \Delta f_*(E \cap T)$ for every set T.

9. For every homeomorphism g, f has a finite or infinite relative derivative f'_g both Δf^*-almost everywhere and Δg^*-almost everywhere in E.

10. f has a derivative, finite or infinite, Δf^*-almost everywhere in E.

Proof. $((1) \Leftrightarrow (2))$. If Δf^* is finite on E then there is a full covering relation β on E so that $\mathrm{Var}(\Delta f, \beta) < +\infty$. The partition of Lemma 2.10 may be used to supply directly the sequence of sets $\{E_i\}$ with the property (2); these are then easily arranged to form a Borel partition of E. If more generally the measure Δf^* is σ-finite on E the above argument applies to each of a sequence of Borel sets covering E. This proves the implication $((1) \Rightarrow (2))$. Conversely if f has uniformly bounded variation on each set $\overline{E}_i$ then certainly Δf^* is finite on E_i and, consequently, Δf^* is σ-finite on E.

$\quad((1) \Leftrightarrow (3))$. If Δf^* is finite on a bounded set E then there are β and $\{E_i\}$ as before; because of Lemma 2.11, the sequence of sets may be taken as increasing. We now construct the following functions: we set $g_{1n}(x) = g_{2n}(x) = f(x)$ for each $x \in \overline{E}_n$ and in the contiguous intervals $\{(a_i, b_i)\}$ to $\overline{E}_n$ we define $g_{1n}(x)$ and $g_{2n}(x)$ in such a way that both functions are continuous, so that

$$f(x) \le g_{1n}(x) \le \sup_{t \in (a_i, b_i)} f(t)$$

and

$$f(x) \geq g_{2n}(x) \geq \inf_{t \in (a_i, b_i)} f(t)$$

and such that the variations of g_{1n} and g_{2n} in this interval do not exceed $\omega f([a_i, b_i])$. From the fact that f has uniformly bounded variation on each set $\overline{E}_i$ it can be shown that $\sum_{i=1}^{\infty} \omega f([a_i, b_i]) < +\infty$. Hence both g_{1n} and g_{2n} have bounded variation on the real line.

Now for each $x \in E_n$ we have $0 = g_{1n}(x) - g_{2n}(x)$ for $x \in E_n$ and in general we have $g_{1n} \geq f \geq g_{2n}$ and $0 \leq g_{1n}(x) - g_{2n}(x)$; on the contiguous intervals $\{(a_i, b_i)\}$ to E_n the oscillation of $g_{1n}(x) - g_{2n}(x)$ is less than $\omega f([a_i, b_i])$. We may deduce from this that

$$V^*(g_{1n} - g_{2n}, E_n) = 0$$

and hence that

$$V^*(g_{1n} - f, E_n) = V^*(g_{2n} - f, E_n) = 0.$$

This establishes $((1) \Rightarrow (3))$ under these hypotheses on E; the extension to unbounded sets on which Δf^* is σ–finite is straightforward. The converse direction, $((3) \Rightarrow (1))$, is immediate since if $\Delta f \equiv \Delta g_n$ on E_n for some sequence of continuous functions $\{g_n\}$ each of bounded variation then, by Theorem 3.25, $\Delta f^*(E_n) = \Delta g_n^*(E_n) < +\infty$.

$((1) \Rightarrow (4))$. The representation in the preceding proof already shows that f is VB$_*$ on each of the sets E_n and so must be VBG$_*$ on E.

$((4) \Leftrightarrow (5))$. This is a theorem of Ward and is proved in Saks [17, pp. 236–237].

$((5) \Rightarrow (6))$. We make the observation that should a function f have everywhere on a set E finite derivates then it is possible to find an increasing homeomorphism g_1 in such a way that $f \circ g_1$ has a finite derivative everywhere on $g^{-1}(E)$. Then if we assume (5) there is a function g_2 so that $\overline{D}(\Delta f, \Delta g_2, x)$ is finite at every point of E; this means $f \circ g_2$ has finite derivates in $g_2^{-1}(E)$ and then by our first observation there is a g_1 so that $f \circ g_2 \circ g_1$ has a finite derivative in $g_1^{-1}(g_2^{-1}(E))$. Thus the function $g = g_2 \circ g_1$ does the task required in (6).

To prove this observation function suppose f has everywhere on a set E finite derivates; let E_1 denote the set of points in E at which the actual derivative fails to exist. This set has zero measure and so there is a superset

E_2 that is a $\mathcal{G}_\delta$ of measure zero. By a lemma of Zahorski (see [2, Lemma 1.2, p. 124]) there is a continuous increasing function k and a positive number c so that $k'(x) = +\infty$ on E_2 and $k'(x) > c$ elsewhere. Take g as the inverse of k and then it is easy to check that $f \circ g$ has a zero derivative in $g^{-1}(E_2 \cap E)$ and a finite derivative in $g^{-1}(E \setminus E_1)$.

$((6) \Rightarrow (1))$. For any set E and any homeomorphism g one has

$$\Delta f \circ g^*(g^{-1}(E)) = \Delta f^*(E).$$

Under the assumptions in (6) the measure $(\Delta f \circ g)^*$ must be σ–finite on $g^{-1}(E)$ and so by applying the homeomorphism g we find that Δf^* must be σ–finite on E.

$((3) \Rightarrow (8))$. The representation in (3) allows us to conclude that $\Delta f^* = \Delta g_n^*$ and $\Delta f_* = \Delta g_{n*}$ on the set E_n. But for continuous functions of bounded variation we know that $\Delta g_n^* = \Delta g_{n*}$ and so $\Delta f^* = \Delta f_*$ on each E_n. This equality must extend to all of $\bigcup E_n$ since the sequence of sets is increasing and so (8) is established.

$((8) \Rightarrow (1))$. Since the measure Δf_* is invariably σ–finite (Theorem 7.7) the identity in assertion (8) can occur only on a set on which Δf^* is σ–finite so that assertion (1) follows.

$((8) \Rightarrow (9))$. We use the identity in assertion (8) together with the σ–finiteness of the measure Δf^* which we have just seen to prove this exactly as in the proof of Theorem 7.1. In this way we can establish that $\overline{D}(\Delta f, \Delta g, x) = \underline{D}(\Delta f, \Delta g, x)$ for Δf^*–almost every and for Δg^*–almost every point in E. Since f and g are continuous this means that the actual relative derivative $f_g'(x)$ (including $\pm\infty$) must exist at these points. It is well known that there can be only countably many such points and so f_g' exists for Δf^*–almost every and for Δg^*–almost every point in E as required to establish (9).

$((9) \Rightarrow (10))$ is immediate.

$((10) \Rightarrow (1))$. If (10) holds then $E = E_1 \cup E_2 \cup E_3$ which are the sets of points in E at which the derivative is finite, is infinite or fails to exist. By hypothesis $\Delta f^*(E_3) = 0$ and by Theorem 4.4 we know that Δf^* is σ–finite on E_1. By [17, Theorem 10.1, p. 234] the function f is VBG$_*$ on E_2 and we know now that this requires Δf^* to be σ–finite on E_2. Thus finally Δf^* is σ–finite on E as required to prove (1).

Finally the proof is completed by showing that (7) is also equivalent. The implication $((7) \Rightarrow (1))$ is immediate. Conversely if we use (5) and (9) we obtain a function g such that the relative derivative $g_1 = f'_g$ exists finitely on $E \setminus N$ for some set N with $\Delta f^*(N) = \Delta g^*(N) = 0$. We define g_1 to be 1 off of N and this gives $\Delta f \equiv g_1 \Delta g$ on $E \setminus N$; but because $\Delta f \equiv g_1 \Delta g \equiv 0$ on N we must have $\Delta f \equiv g_1 \Delta g$ on E as required. This completes the proof.

From this theorem we may also derive the following characterizations of zero variation, finite variation and σ–finite variation.

THEOREM 7.9 *Let f be a continuous function and E a bounded Borel set. Then*

1. *$\Delta f^*(E) = 0$ if and only if for some continuous, monotonic function g the function $(f \circ g)'$ has a zero derivative on the set $g^{-1}(E)$.*

2. *$\Delta f^*(E) < \infty$ if and only if for some continuous, monotonic function g the function $(f \circ g)'$ has a bounded derivative on the set $g^{-1}(E)$.*

3. *Δf^* is σ–finite on E if and only if for some continuous, monotonic function g the function $(f \circ g)'$ has a finite derivative on the set $g^{-1}(E)$.*

Proof. We prove the second assertion only. The first is similarly proved and the third is contained in the preceding theorem. Suppose that $\Delta f^*(E) < \infty$. Choose a full covering relation β on E so that $\text{Var}(\Delta f, \beta) < \infty$ and define the function $k(x) = \text{Var}(\Delta f, \beta((-\infty, x])) + x$. Note that k is strictly increasing (but not necessarily continuous) and that because of the elementary inequality

$$\text{Var}(\Delta f, \beta((-\infty, x])) + \text{Var}(\Delta f, \beta([x, y])) \leq \text{Var}(\Delta f, \beta((-\infty, y]))$$

we have $|f(y) - f(x)| \leq k(y) - k(x)$ for any pair $([x, y], x_0) \in \beta$. Since f is continuous and k is monotone this inequality extends to $|f(y) - f(x)| \leq k(y-) - k(x+)$ for any pair $([x, y], x_0) \in \beta$.

Choose a continuous, nondecreasing function g so that $g(k(x)) = x$ for all x. For any point $s_0 \in g^{-1}(E)$ there is a point $x_0 \in E$ with $g(s_0) = x_0$ and a positive δ so that, whenever $x < x_0 < y$ and $y - x < \delta$, $([x, y], x_0) \in \beta$. As g is continuous we may choose $\delta' > 0$ so that $s < s_0 < t$, $t - s < \delta'$ requires that $g(t) - g(s) < \delta$.

If $s_0 \in g^{-1}(E)$, $s < s_0 < t$ and $t - s < \delta'$ then we can select points $x \leq x_0 \leq y$ so that $x_0 \in E$, $g(s_0) = x_0$, $f(g(t)) = f(y)$, $f(g(s)) = f(x)$,

$t \geq g(y-)$, $s \leq g(x+)$ and hence, $y - x = g(t) - g(s) < \delta$. This means that $([x,y], x_0) \in \beta$ and we must have for $x \neq y$ the inequalities

$$\left| \frac{f(g(t)) - f(g(s))}{t - s} \right| \leq \left| \frac{f(y) - f(x)}{g(y-) - g(x+)} \right| \leq 1.$$

This shows that $f \circ g$ has bounded derivates on the set $g^{-1}(E)$. As in the proof of Theorem 7.8 a further change of variable can be used to produce the required characterization.

Conversely for (2) again if $|(f \circ g)'(s)| < M$ on the set $g^{-1}(E)$ let β be the collection of all pairs $([x,y], x_0)$ with $x_0 \in E$, $x < x_0 < y$ such that, for some $A < s < t < B$, $g(s) = x$, $g(t) = y$ and $|f(g(t)) - f(g(s))| \leq M|t - s|$ where A and B are bounds for $g^{-1}(E)$. Because $|(f \circ g)'(s)| < M$ we check that β is a full covering relation on E and evidently $\mathrm{Var}(\Delta f, \beta) \leq M(B - A) < \infty$ so that $\Delta f^*(E) < \infty$ as required.

7.3 Absolute continuity and singularity

In this section we show how the notions of absolute continuity and singularity may be lifted to functions with finite or σ–finite variation. We use the classical notions of AC_* and ACG_* (as defined in Saks [17]); in addition we say that a function f is *uniformly absolutely continuous* on a set E if for any $\epsilon > 0$ there is a positive numbers δ so that the sums $\sum_{i=1}^{n} |\Delta f(I)|$ are smaller than ϵ for any sequence $\{I_1, I_2, \ldots, I_n\}$ of nonoverlapping intervals of total length less than δ and each containing a point of E as an interior point.

We state without proof some characterizations of ACG_* functions; the proof is mostly similar to that of Theorem 7.8 and is in any case duplicated in part by the more general Theorem 7.17.

THEOREM 7.10 *Let f be a continuous function and let E be a Borel set of real numbers. Then the following assertions are equivalent:*

1. *Δf^* is σ–finite on E and $\Delta f^*(N \cap E) = 0$ for every set N of Lebesgue measure zero.*

2. *E has a Borel partition $E = \bigcup_{i=1}^{\infty} E_n$ and $\Delta f \ll \ell$ on each E_n.*

3. *E has a Borel partition $E = \bigcup_{i=1}^{\infty} E_n$ such that f is uniformly absolutely continuous on each set $\overline{E_i}$.*

4. E may be written as an increasing union $\bigcup_{i=1}^{+\infty} E_i$ of Borel sets in such a way that, for each n, $\Delta f \equiv \Delta g_n$ on E_n for some sequence of absolutely continuous functions $\{g_n\}$.

5. f is ACG_* on E.

6. f is VBG_* on E and $f[N \cap E]$ has measure zero for every set N of Lebesgue measure zero.

7. f has a finite derivative Δf^*-almost everywhere in E.

8. For every Lebesgue measurable subset E' of E,

$$\Delta f^*(E') = \int_{E'} |f'(x)| \, dx$$

(where $f'(x)$ is taken as zero at the points where it fails to exist).

We also have as a corollary the following extension of the de la Vallée Poussin theorem (Theorem 7.4).

COROLLARY 7.11 *Suppose that f is a continuous function that has finite variation on a Borel set E. Then $f'(x)$ exists a.e. on E and*

$$\Delta f^*(E) = (L) \int_E |f'(x)| \, dx + \Delta f^* \left(\{x \in E : f'(x) = \pm\infty\} \right).$$

Moreover

$$\Delta f^*(E) = (L) \int_E |f'(x)| \, dx$$

if and only if f is ACG_* on E.

In contrast to this statement we present the following result for the fine variational measure; note that here the decomposition is available for an arbitrary continuous function without any assumptions on the variation.

COROLLARY 7.12 *Let f be a continuous function and E be a Borel set; then*

$$\Delta f_*(E) = (L) \int_{E \cap D} |f'(x)| \, dx + \Delta f_* \left(\{x \in E : f'(x) = \pm\infty\} \right)$$

where D is the set of points at which f has a finite derivative.

Proof. Let B_1 be the set of points at which f has a zero derived number, let B_2 be the set of points x at which $f'(x) = \pm\infty$, let B_3 be the set of points x at which $\overline{D}f(x) < \infty$ or $\underline{D}f(x) > -\infty$ and let B_4 be the set of points x at which $-\infty < D_+f(x) \leq D^+f(x) < \infty$ or $-\infty < D_-f(x) \leq D^-f(x) < \infty$. As f is continuous this exhausts all possibilities with at most countably many exceptions. It is easy to check that $\Delta f_*(B_1) = 0$. On each of the sets B_2, B_3 and B_4 the function f has σ–finite variation (see [17, pp. 234–235]). Since these are evidently Borel sets Corollary 7.11 may be applied to give the present decomposition.

A further corollary to this might be mentioned here; this too may be used to obtain a criterion for a function to be absolutely continuous. A proof may be obtained from [17, Theorem 6.6, p. 280] and [17, Theorem 7.3, p. 284].

COROLLARY 7.13 *Let f be a continuous function that satisfies Lusin's condition (N) on an interval $[a, b]$. Then*

$$|f(b) - f(a)| \leq \Delta f_*(D \cap (a, b))$$

where D is the set of points at which f has a finite nonzero derivative.

For functions with finite variation a further characterization of the notion of absolute continuity is available.

THEOREM 7.14 *Let f be a continuous function and E a bounded Borel set. Then the following are equivalent:*

1. *$\Delta f \ll \ell$ on E.*

2. *$\Delta f \ll \ell$ weakly on E and Δf^* is σ–finite on E.*

3. *Δf^* is finite on E and $\Delta f^*(N \cap E) = 0$ for every set N of Lebesgue measure zero.*

Proof. $((1) \Rightarrow (2))$ is immediate. For $(2) \Rightarrow (3)$ choose a $\delta > 0$ and a fine covering relation β on E so that $\sum_\pi |I| < \delta$ implies that $\sum_\pi |\Delta f(I)| < 1$. Let $G_1, G_2, \ldots G_N$ be a covering of E by open sets of diameter smaller than δ. Then

$$\Delta f^*(E) = \Delta f_*(E) \leq V^*(\Delta f, \beta) \leq \sum_{i=1}^{N} V^*(\Delta f, \beta(G_i)) \leq N.$$

This shows that $\Delta f^*(E)$ is finite and the measure-theoretic absolute continuity part of (3) follows directly from the discussion in Section 5.3. Finally for (3) $\Rightarrow$ (1) note that, because of Theorem 7.8, f' must exist finitely everywhere in E except on a set N with $\Delta f^*(N) = |N| = 0$. From this we deduce $\Delta f \equiv f'\ell$ on E; by Theorem 5.8 we obtain (1) and the proof is complete.

We turn now to a collection of characterizations of singular functions. A monotonic function f is commonly called singular if it has a zero derivative almost everywhere. The language of singular functions can be expanded by taking any of the following equivalent statements as a definition. Note that we require finite variation for the function on the given set.

THEOREM 7.15 *Let f be a continuous function and E a bounded Borel set on which f has finite variation. The following assertions are equivalent:*

1. $\Delta f \perp \ell$ *on E.*

2. $\Delta f \perp \ell$ *weakly on E.*

3. $f'(x) = 0$ *almost everywhere on E.*

4. $f'(x) = \pm\infty$ Δf^*-*almost everywhere on E.*

5. *There is a subset $N \subset E$ such that $\Delta f^*(E \setminus N) = |N| = 0$.*

6. $\sqrt{|\Delta f|\ell} \equiv 0$ *on E.*

7. *For every $\delta > 0$ there are nonoverlapping intervals $I_1, I_2, \ldots I_m$ so that $\sum_{k=1}^m |I_k| < \delta$ and $\Delta f^*\left(E \setminus \bigcup_{k=1}^m I_k\right) < \delta$.*

8. *For every $\delta > 0$ there are nonoverlapping intervals $I_1, I_2, \ldots I_m$ so that each $|I_k| < \delta$, $\sum_{k=1}^m |\Delta f(I_k)| < \delta$ and $|E \setminus \bigcup_{k=1}^m I_k| < \delta$.*

The proof is an application of the general material in Section 5.6 and is omitted. A similar set of characterizations would be available for the mutual singularity of a pair of functions f and g under assumptions on their variation.

The final theorem in this section may be considered as an application of the characterizations of singularity and absolute continuity to an arc length problem. The expression $\Gamma_f^*(E)$, while evidently related to arc length, depends on more than just the graph $\{(x, y) : f(x) = y, x \in E\}$. The proof is omitted.

THEOREM 7.16 *Let f be a continuous function, E a Borel set on which f has finite variation and Γ_f the interval function $\sqrt{\ell^2 + (\Delta F)^2}$. If f is absolutely continuous on E then*

$$\Gamma_f^*(E) = \int_E \sqrt{(f'(x))^2 + 1}\, dx.$$

If f is singular on E then

$$\Gamma_f^*(E) = \Delta f^*(E) + |E|.$$

7.4 s–absolute continuity

In this section we apply our general results to obtaining a characterization of the functions which are, in a sense made precise below, absolutely continuous with respect to the Hausdorff s–dimensional measure or to the s–dimensional packing measure. This can be considered an extension of Rogers and Taylor [16]; see also the exposition in Rogers [15].

Let f be a real function and E a set of real numbers and suppose that $0 < s \leq 1$; we say that f is *s–absolutely continuous* on the set E provided that, in the language of Section 5.1,

$$\Delta f \ll \ell^s \text{ on } E.$$

Expressed explicitly this says that for every $\epsilon > 0$ there are a $\delta > 0$ and a full covering relation β on E so that whenever

$$(I_1, x_1), (I_2, x_2), (I_3, x_4), \ldots (I_n, x_n)$$

is a packing from β for which

$$\sum_{i=1}^{n} |I_i|^s < \delta$$

then

$$\sum_{i=1}^{n} |\Delta f(I_i)| < \epsilon.$$

We give a characterization of this notion in terms of the measure μ_s, and in terms of generalized Lipschitz conditions. For convenience we say that f

is *uniformly s–Lipschitz* on a set E if Δf is $\mathrm{Lip}(\ell^s)$ on E. Again this says that for some positive δ an inequality

$$|f(d) - f(c)| \le C(d-c)^s$$

holds for any interval (c,d) that contains a point of E and has length less than δ. An integral taken with respect to the Hausdorff s–dimensional measure λ^s will be written

$$\int_B f(x)\,dx^s$$

following Besicovitch [1]. We know from Theorem 3.23 that

$$\int_B f(x)\,dx^s = (f\ell^s)_*(B)$$

for Borel sets B and nonnegative Borel functions f which provides us with a useful representation of this integral. As mentioned in the introduction this allows us to interpret this integral as a limit of some sequence of Riemann sums of the form $\sum f(\xi_i)(x_{i+1} - x_i)^s$.

THEOREM 7.17 *Let f be a continuous real function that has finite variation on a Borel set E. Then the following assertions are equivalent:*

1. *f is s–absolutely continuous on E.*

2. *There is a Borel partition $E = \bigcup_{n=0}^\infty E_n$ of E such that $\Delta f^*(E_0) = 0$ and f is uniformly s–Lipschitz on each set $\overline{E}_n$ $(n \ge 1)$.*

3. *$\overline{D}_s(f,x) < +\infty$ at Δf^*–almost every point x in E.*

4. *$\Delta f^*(E \cap N) = 0$ for every set N with $\mu_s(N) = 0$.*

5. *For every Borel set $B \subset E$,*

$$\Delta f^*(B) = \Delta f^*(B \cap E_0) + (L)\int_{B\cap E_+} \overline{D}_s(f,x)\,dx^s$$

   where

$$E_0 = \left\{ x \in E : \overline{D}_s(f,x) = 0 \right\}$$

   and

$$E_+ = \left\{ x \in E : 0 < \overline{D}_s(f,x) < +\infty \right\}.$$

Proof. If item (5) is true and $\mu_s(N) = 0$ then the representation in (5) shows that $\Delta f^*(N \cap E_+) = 0$. Theorem 4.13 shows then that $\Delta f^*(N \cap E_0) = 0$ and so finally $\Delta f^*(E \cap N) = 0$ as required to prove (4).

If item (4) is true define the set

$$A_0 = \left\{ x \in E : \overline{D}\left(\Delta f, \ell^s, x\right) = +\infty \right\}.$$

By Theorem 4.3, since Δf is σ–finite, we have $\mu_s(A_0) = 0$. By the assumptions in (4) this means that $\Delta f^*(A_0) = 0$ as required for (3).

If (3) is true then we may produce the partition in item (2). Set A_0 as above (so that $\Delta f^*(A_0) = 0$) and apply Theorem 4.12 to the set $E \setminus A_0$ to obtain a partition for which f is uniformly $\text{Lip}(\ell^s)$ on each member. This is precisely the statement in item (2).

If (2) holds then certainly $\Delta f \ll \ell^s$ on each member of the sequence $\{E_i\}$. By Theorem 5.7 we have $\Delta f \ll \ell^s$ on E, proving item (1).

If (1) holds and $\mu_s(T) = 0$ then, as before, $\Delta f_*(E \cap T) = 0$. Since f is VBG_* on E the measures Δf_* and Δf^* agree and so $\Delta f^*(E \cap T) = 0$ as required to establish (4).

Finally then, since we now have the equivalence of (1)–(4), we may show that (5) follows from (3). If (3) holds then certainly for any Borel set $B \subset E$

$$\Delta f^*(B) = \Delta f^*(B \cap E_0) + \Delta f^*(B \cap E_+).$$

The representation

$$\Delta f^*(B \cap E_+) = (L) \int_{B \cap E_+} \overline{D}_s(f, x)\, dx^s$$

is a consequence of Theorems 6.4 and 3.23. This establishes (5) and the proof is complete.

The absolute continuity of the measure Δf^* relative to the full measure μ^s (rather than merely the fine measure μ_s) is an apparently weaker condition. A version of Theorem 7.17 that characterizes this could be formulated on the basis of Theorem 5.20.

Finally from Theorem 7.17 we may deduce the series representation of s–absolutely continuous functions given by Rogers [15].

COROLLARY 7.18 *Let f be a continuous, nondecreasing real function on the interval $[0, 1]$ that is s–absolutely continuous on $[0, 1]$. Then there is a* *sequence $\{g_n\}$ of continuous nondecreasing real functions on the interval $[0,1]$ so that each g_n is uniformly s–Lipschitz on $[0,1]$ and, for every $0 \leq t \leq 1$,*

$$f(t) = \sum_{n=1}^{\infty} g_n(t).$$

Proof.   We suppose that f is defined on the entire real line by setting $f(x) = f(0)$ for $x < 0$ and $f(x) = f(1)$ for $x > 1$. We apply Theorem 7.17 to obtain

$$f(t) - f(0) = \Delta f_\bullet^*(E_0) + \sum_{k=1}^{\infty} \Delta f^*(E_k \cap [0,t])$$

where $\{E_k\}$ is the Borel partition of $[0,1]$ promised in that theorem. We have then $\Delta f^*(E_0) = 0$ and f is uniformly s–Lipschitz on each $\overline{E}_k$ for $k \geq 1$.

Note that $\Delta f^*(E_k \cap J) \leq \Delta f(J)$ for any subinterval J of I and consequently if we set

$$g_k(t) = \Delta f^*(E_k \cap [0,t])$$

then g_k is uniformly s–Lipschitz on $\overline{E}_k$; but Δg_k vanishes on any interval J with $J^0 \cap E_k = \emptyset$ and so in fact g_k is uniformly s–Lipschitz on all of $[0,1]$. Since this is the representation of the theorem the proof is complete.

Let us conclude our discussion of s–absolute continuity by sketching out a possible application. Suppose that we are given a continuous function f for which it is known that the derivative $f'(x)$ exists everywhere but at the points of a set N. What additional information is required in order that we may claim that

$$f(b) - f(a) = \int_a^b f'(t)\, dt \tag{32}$$

holds for all intervals $[a,b]$? If N is countable then (32) holds in the Denjoy-Perron sense. (If we know that f' is Lebesgue integrable or is bounded on one side then the integral may be interpreted in the Lebesgue sense.) If we know only that $|N| = 0$ then again (32) holds if f is absolutely continuous.

Suppose now that we have information between these extremes, namely that N has Hausdorff dimension less than s. The results of this section would allow us to prove that once again (32) holds if f is s–absolutely continuous and has σ–finite variation on N.

7.5 s–singular functions

In this section we apply our general results to obtaining a characterization of the functions which are singular with respect to the Hausdorff s–dimensional measure or to the s–dimensional packing measure. As before this can be considered an extension of some related results in [15] and [16].

We say that a real function f is s–singular on a set E if $\Delta f \perp \ell^s$ on E and weakly s–singular on E if $\Delta f \perp \ell^s$ weakly on E. Our theorems characterize these notions for functions that are VBG$_*$. For the most part they follow immediately from the material in Section 5.6 by using the density property, that $\Delta f^* = \Delta f_*$ on any set on which f is VBG$_*$.

THEOREM 7.19 *Let f be a continuous real function that is VBG$_*$ on a Borel set E. Then the following assertions are equivalent.*

1. *f is s–singular on E .*

2. *For Δf^*–almost every x in E the derivative $\underline{D}_s(f,x) = +\infty$ and for μ^s–almost every x in E the derivative $\underline{D}_s(f,x) = 0$.*

3. *There is a subset N of E so that $\Delta f^*(E \setminus N) = \mu^s(N) = 0$.*

4. *μ^s is concentrated on the set where $\underline{D}_s(f,x)$ is zero and Δf^* is concentrated on the set where $\underline{D}_s(f,x)$ is infinite.*

Proof. We apply Theorem 5.20 with $h = \Delta f$ and $k = \ell^s$ and this supplies most of the proof. The only part of the proof that needs to be provided is the additional statement in (2) that for μ^s–almost every x in E the derivative $\underline{D}_s(f,x) = 0$. Since the measure Δf^* is σ–finite on E the set of points in E at which the derivate $\underline{D}_s(f,x)$ is positive must have σ–finite μ^s–measure too; this follows from Theorem 4.4 for example. But we know from Theorem 6.5 that the measure μ^s must then vanish. This completes the assertion in (2). Finally then the assertion in (4) is just a rewording of this and the proof is complete.

Our final theorem characterizes the weak singularity notion in this setting.

THEOREM 7.20 *Let f be a continuous real function that has finite variation on a Borel set E. Then the following assertions are equivalent.*

1. f *is weakly s–singular on* E .

2. *For μ_s–almost every x in E the derivative $\overline{D}_s(f,x) = 0$ and for Δf^*–almost every x in E the derivative $\overline{D}_s(f,x) = +\infty$.*

3. *There is a subset N of E so that $\Delta f^*(E \setminus N) = \mu_s(N) = 0$.*

4. μ_s *is concentrated on the set where $\overline{D}_s(f,x)$ is zero and Δf^* is concentrated on the set where $\overline{D}_s(f,x)$ is infinite.*

5. *for every $\delta > 0$ there is a finite sequence of nonoverlapping intervals $\{I_i\}$ so that $\sum |I_i|^s < \delta$ and $\Delta^* f(E \setminus \bigcup I_i) < \delta$.*

Proof.   In much the same way the preceding theorem has been proved we apply Theorem 5.21 with $h = \Delta f$ and $k = \ell^s$. Because of the density property $\Delta f^* = \Delta f_*$ on E this supplies most of the proof. The only part of the proof not already clear is the additional statement in (2) that for μ_s–almost every x in E the derivative $\overline{D}_s(f,x) = 0$. We have that the set of points in E at which the derivate $\overline{D}_s(f,x)$ is positive and finite has Δf^*–measure zero. From Corollary 4.7 we obtain that this set must have zero μ_s–measure too. Since we already know that $\overline{D}_s(f,x)$ is finite μ_s–almost everywhere in E this completes the assertion in (2). Finally then the assertion in (4) is just a rewording of this and the proof is complete.

References

[1] A. S. Besicovitch. On linear points of fractional dimension. *Math. Annalen*, 101:161–193, 1929.

[2] A. M. Bruckner. *Differentiation of Real Functions. Lecture Notes in Mathematics 659*, Springer-Verlag, Berlin, 1978.

[3] M. de Guzmán. A general form of the Vitali theorem. *Colloq. Math.*, 34:69–72, 1975.

[4] K. J. Falconer. *The Geometry of Fractal Sets*. Cambridge University Press, Cambridge, 1985.

[5] H. Federer. *Geometric Measure Theory*. Springer, New York, 1969.

[6] C. A. Hayes and C. Y. Pauc. *Derivation and Martingales.* Springer-Verlag, Berlin, 1970.

[7] R. Henstock. Generalized integrals of vector-valued functions. *Proc. London Math. Soc.*, (3) 19:317–344, 1969.

[8] J. P. Kahane and R. Salem. *Ensembles Parfait et séries trigonométriques.* Hermann, Paris, 1963.

[9] H. Kober. On singular functions of bounded variation. *J. London Math. Soc.*, 23:222–229, 1948.

[10] A. Kolmogorov. Üntersuchen über der integralbegriffe. *Math. Annalen*, 103:654–696, 1930.

[11] J. Kurzweil. Generalized ordinary differential equations and continuous dependence on a parameter. *Czech. Math. Journal*, 82:418–449, 1957.

[12] S. Leader. A concept of differential based on variational equivalence under generalized Riemann integration. *Real Analysis Exchange*, 12:144–175, 1986.

[13] S. Leader. What is a differential? A new answer from the generalized Riemann integral. *American Math. Monthly*, 93:348–356, 1986.

[14] J. G. Mauldon. Continuous functions with zero derivative almost everywhere. *Quart. J. Math. Oxford*, (2) 17:256–262, 1966.

[15] C. A. Rogers. *Hausdorff Measures.* Cambridge University Press, Cambridge, 1970.

[16] C. A. Rogers and S. J. Taylor. Functions continuous and singular with repect to Hausdorff measure. *Mathematika*, 8:1–31, 1961.

[17] S. Saks. *Theory of the Integral.* Monografie Matematyczne 7, Warsaw, 1937.

[18] S. J. Taylor and C. Tricot. Packing measure and its evaluation for a Brownian path. *Trans. American Math. Soc.*, 288:679–699, 1985.

[19] S. J. Taylor and C. Tricot. The packing measure of rectifiable subsets of the plane. *Math. Proc. Cambridge Phil. Soc.*, 99:285–296, 1986.

[20] G. Temple. *The Structure of Lebesgue Integration Theory.* Oxford Univ. Press, 1971.

[21] B. S. Thomson. Derivation bases on the real line. I and II. *Real Analysis Exchange*, 8:67–207 and 278–442, 1982–83.

[22] B. S. Thomson. *Real Functions. Lecture Notes in Mathematics 1170*, Springer-Verlag, Berlin, 1985.

[23] C. Tricot. Two definitions of fractional dimension. *Proceedings Cambridge Philosohical Society*, 91:57–74, 1982.

[24] A. I. Tulcea and C. I. Tulcea. *Topics in the Theory of Lifting.* Springer-Verlag, New York, 1969.

Index

MEMOIRS of the American Mathematical Society

SUBMISSION. This journal is designed particularly for long research papers (and groups of cognate papers) in pure and applied mathematics. The papers, in general, are longer than those in the TRANSACTIONS of the American Mathematical Society, with which it shares an editorial committee. Mathematical papers intended for publication in the Memoirs should be addressed to one of the editors:

Ordinary differential equations, partial differential equations and applied mathematics to ROGER D. NUSSBAUM, Department of Mathematics, Rutgers University, New Brunswick, NJ 08903

Harmonic analysis, representation theory and Lie theory to AVNER D. ASH, Department of Mathematics, The Ohio State University, 231 West 18th Avenue, Columbus, OH 43210

Abstract analysis to MASAMICHI TAKESAKI, Department of Mathematics, University of California, Los Angeles, CA 90024

Real and harmonic analysis to DAVID JERISON, Department of Mathematics, M.I.T., Rm 2–180, Cambridge, MA 02139

Algebra and algebraic geometry to JUDITH D. SALLY, Department of Mathematics, Northwestern University, Evanston, IL 60208

Geometric topology and general topology to JAMES W. CANNON, Department of Mathematics, Brigham Young University, Provo, UT 84602

Algebraic topology and differential topology to RALPH COHEN, Department of Mathematics, Stanford University, Stanford, CA 94305

Global analysis and differential geometry to JERRY L. KAZDAN, Department of Mathematics, University of Pennsylvania, E1, Philadelphia, PA 19104-6395

Probability and statistics to RICHARD DURRETT, Department of Mathematics, Cornell University, Ithaca, NY 14853-7901

Combinatorics and number theory to CARL POMERANCE, Department of Mathematics, University of Georgia, Athens, GA 30602

Logic, set theory, general topology and universal algebra to JAMES E. BAUMGARTNER, Department of Mathematics, Dartmouth College, Hanover, NH 03755

Algebraic number theory, analytic number theory and modular forms to AUDREY TERRAS, Department of Mathematics, University of California at San Diego, La Jolla, CA 92093

Complex analysis and nonlinear partial differential equations to SUN-YUNG A. CHANG, Department of Mathematics, University of California at Los Angeles, Los Angeles, CA 90024

All other communications to the editors should be addressed to the Managing Editor, DAVID J. SALTMAN, Department of Mathematics, University of Texas at Austin, Austin, TX 78713.

General instructions to authors for

PREPARING REPRODUCTION COPY FOR MEMOIRS

> For more detailed instructions send for AMS booklet, "A Guide for Authors of Memoirs."
> Write to Editorial Offices, American Mathematical Society, P.O. Box 6248,
> Providence, R.I. 02940.

MEMOIRS are printed by photo-offset from camera copy fully prepared by the author. This means that the finished book will look exactly like the copy submitted. Thus the author will want to use a good quality typewriter with a new, medium-inked black ribbon, and submit clean copy on the appropriate model paper.

Model Paper, provided at no cost by the AMS, is paper marked with blue lines that confine the copy to the appropriate size.

Special Characters may be filled in carefully freehand, using dense black ink, or **INSTANT** ("rub-on") **LETTERING** may be used. These may be available at a local art supply store.

Diagrams may be drawn in black ink either directly on the model sheet, or on a separate sheet and pasted with rubber cement into spaces left for them in the text. Ballpoint pen is not acceptable.

Page Headings (Running Heads) should be centered, in CAPITAL LETTERS (preferably), at the top of the page — just above the blue line and touching it.

LEFT-hand, EVEN-numbered pages should be headed with the AUTHOR'S NAME;

RIGHT-hand, ODD-numbered pages should be headed with the TITLE of the paper (in shortened form if necessary).

Exceptions: PAGE 1 and any other page that carries a display title require NO RUNNING HEADS.

Page Numbers should be at the top of the page, on the same line with the running heads.

LEFT-hand, EVEN numbers — flush with left margin;

RIGHT-hand, ODD numbers — flush with right margin.

Exceptions: PAGE 1 and any other page that carries a display title should have page number, centered below the text, on blue line provided.

FRONT MATTER PAGES should be numbered with Roman numerals (lower case), positioned below text in same manner as described above.

MEMOIRS FORMAT

> It is suggested that the material be arranged in pages as indicated below.
> Note: <u>Starred items (*)</u> are requirements of publication.

Front Matter (first pages in book, preceding main body of text).

Page i — *Title, *Author's name.

Page iii — Table of contents.

Page iv — *Abstract (at least 1 sentence and at most 300 words).

Key words and phrases, if desired. (A list which covers the content of the paper adequately enough to be useful for an information retrieval system.)

*<u>1991 Mathematics Subject Classification.</u> This classification represents the primary and secondary subjects of the paper, and the scheme can be found in Annual Subject Indexes of MATHEMATICAL REVIEWS beginnning in 1990.

Page 1 — Preface, introduction, or any other matter not belonging in body of text.

Footnotes: *Received by the editor date.
Support information — grants, credits, etc.

First Page Following Introduction – Chapter Title (dropped 1 inch from top line, and centered). Beginning of Text.

Last Page (at bottom) – Author's affiliation.